GOVERNING CITIES ON THE MOVE

Governing Cities on the Move

Functional and management perspectives on
transformations of European urban infrastructures

Edited by
MARTIN DIJST
Utrecht University, The Netherlands

WALTER SCHENKEL
University of Zürich, Switzerland

ISABELLE THOMAS
Université Catholique de Louvain, Louvain-la-Neuve, Belgium

Routledge
Taylor & Francis Group

LONDON AND NEW YORK

First published 2002 by Ashgate Publishing

Reissued 2018 by Routledge
2 Park Square, Milton Park, Abingdon, Oxon OX14 4RN
711 Third Avenue, New York, NY 10017, USA

Routledge is an imprint of the Taylor & Francis Group, an informa business

A Library of Congress record exists under LC control number: 2001095430

ISBN 13: 978-1-138-72565-2 (hbk)
ISBN 13: 978-1-138-72561-4 (pbk)
ISBN 13: 978-1-315-19178-2 (ebk)

Contents

Conclusions

List of Figures

List of Tables

List of Contributors

Boffi, Mario
He works at the Department of Sociology and Social Research of the University of Milan – Bicocca, Italy. Boffi is professor of Data Base Management at the Faculty of Statistics and he is involved in several projects on urban sociological data analysis.

Dijst, Martin
He is an associate professor of urban geography at the Urban Research centre Utrecht (URU), Faculty of Geographical Sciences, Utrecht University in the Netherlands. His research activities are focused on transportation studies. He is particularly interested in the relations between urbanisation, infrastructure, and the activity/travel patterns of specific population categories, accessibility modelling and multimodal transportation. In 1995 he completed his Ph.D. dissertation entitled "The elliptical life", in which he treats action space as an integral measure for accessibility and mobility.

Güller, Peter
He is an architect SFIT, Planner. He is the owner of Synergo, a private consultancy in Zurich, Switzerland. Main work fields: regional and urban development, transport policy and traffic planning, and environmental issues. He works as a consultant, project manager, researcher and research manager in Switzerland, the European Union and several of its member countries and South-East Asia (Himalaya regions).

Halvorsen, Knut
He is an economist who has been working for about ten years as a researcher at the Norwegian Institute of Urban and Regional

Research in Oslo, Norway. He has mostly concentrated on the conditions for industrial change, both in urban and rural areas, and the last few years he has been interested in new forms for planning and economic organisation. He now works as a managing director for an urban industrial development agency in Oslo, the Oslo Business Region, while holding a part time position as an associate professor at the Departement of Innovation and Economic Organisation at the Norwegian School of Management - BI in Oslo.

Jayet, Hubert
He is Professor of Economics at MEDEE, Department of Economics, Lille University of Science and Technology. His main research themes are urban and regional economics, local labour markets, land and housing prices, local labour markets and migration, local public economics and fiscal competition and spatial econometrics.

Kanaroglou, Pavlos
He is Professor of Geography and Head of the Department of Geography at the University of the Aegean in Greece. He is currently on leave from the Department of Geography at McMaster University, Hamilton, Ontario, Canada. His interests include integrated urban transport and land-use models, travel demand analysis, population analysis and forecasting, urban air pollution and sustainable urban form, and quantitative methods in geography.

Lami, Isabella
She is an Architect and has a Ph.D. in Urban Planning. She is now a researcher in the Faculty of Architecture - Polytechnic of Turin, Italy. Lami is involved in the study of financial analysis use in urban policies, in particular in the major infrastructural transformation of the city, as an instrument to verify the prediction feasibility of plans, and to provide a base for negotiation between public and private representatives.

Mennola, Erkki
He is Doctor in Administrative Sciences with a Ph.D. in Laws and Docent in Local Governance at the University of Tampere, Finland.

He has been working as associate professor in Local Governance and as an independent consultant and adviser in local and regional governance and European regional policy. He has a wide experience of the local and regional administration in Finland, too, as the secretary of metro planning commission, town clerk and manager of regional council. His particular interest is European comparative research of local and regional governance.

Nuvolati, Giampaolo
He works at the Department of Sociology and Social Research, University of Milan - Bicocca, Italy. Nuvolati has worked for many years on social indicators and on quality of life in the cities. More recently he has been involved in several national research projects on Mobility and Urban Time.

Riganti, Paolo
He has a Ph.D. in Urban Planning and real estate market at the Politechnico of Turin. He works at the Politechnico of Milan, Dipartimento Scienze del Territorio. His main research theme is the impact of land-use politics on travel demand.

Schenkel, Walter
He studied history, political science, and philosophy at the University of Zurich. His Master thesis of 1990 was on urban planning processes. He is research fellow at the Zurich Institute of Political Science in 1996-1997 and was guest researcher at the Erasmus University, Rotterdam. He gained his Doctorate in 1998. He is Partner in a private company (Muri&Partner). His main research interests are environmental policy, urban planning, federalism, and network management.

Scott, Darren
He is an assistant professor in the Department of Geography and Geosciences at the University of Louisville, Kentucky, USA. He is also a member of the Logistics and Distribution Institute at this institution. His research interests include activity-based travel

demand modelling, integrated urban transportation and land-use models, microsimulation, object orientation and intermodal transport.

Stein, Véronique
She is a geographer working as a teaching and research assistant at the University of Geneva in Switzerland. Her main interest is in issues of urban and social geography. For the last two years, she has been working on a Swiss Cost-Civitas project on public space planning, in collaboration with the Institute for Research on the Built Environment.

Thomas, Isabelle
She currently has a permanent research position at the Belgian National Fund for Scientific Research (F.N.R.S.), gives lectures at the Department of Geography of the U.C.L. in Louvain-la-Neuve, Belgium and is invited member at the Centre of Operation Research and Econometrics (C.O.R.E.). She has two main research interests: the optimal location-allocation models of human activities and especially the sensitivity of the results of those models to their inputs and their applications to various regional or urban areas, and spatial data analysis (exploratory data analysis, mapping techniques, etc.).

Verhetsel, Ann
She gained her Doctorate in Geography in 1989 and is currently professor at the University of Antwerp in the Faculty of Applied Economics. She teaches courses on Economic Geography, Regional Economics and Human Geography. She is also active in the education of spatial planners. Her research concerns the interaction of geography, economy and spatial planning. Recent research projects were in the domains of sustainable mobility, regional development and geomarketing.

Preface

This is one of three volumes to be published as a result of the action COST A9 *Civitas – Transformation of European Cities and Urban Governance*. This action was launched in 1995, at a time when urban studies had not yet acquired *droit de cité* in the European Union programmes. This initiative was launched by a number of European researchers. It coincided with the emergence of the urban question, which now pervades many European Union programmes.

Here, there is no need to dwell on Europe's obvious urban characteristics. Nor is it necessary to defend the diversity of the European programmes targeted to parts of the city. It suffices to cite two documents published recently by the European Commission. To some extent, these memorandums reflect the centrality of this set of themes. Notably, the EU does not assert any particular competence in this field.

The first document is entitled *Sustainable Urban Development in the European Union: A Framework for Action* and was published in 1999. It stresses four topics which are central to the Civitas co-operation project: the economy, society, the environment, and urban governance. The second document is more pertinent to the work presented in this book. It relates directly to the 5th Framework Programme. That EU initiative explicitly includes the urban dimension in its 'priority action 4' document entitled *City of Tomorrow and the Cultural Heritage.*

The COST framework is a European mechanism to provide scientific and technical assistance for national research programmes. An essential aspect of that mechanism is to facilitate such initiatives. It covers a wide range of fields and includes about thirty European countries today. The COST framework can only exist by virtue of a bottom-up dynamic, which calls for initiatives by researchers of the

Member States. Its work consists of mechanisms of exchange and co-operation.

The COST framework has supported the emergence of this set of urban themes and urban governance on a European scale. At the same time, the co-operative project has allowed researchers and research institutions working on these topics in various countries to meet in an atmosphere conducive to the development of fruitful research in the future. That future is already blending with the present. The Civitas action has made it possible to build upon the research done as part of the programme on urban development and urban governance. That research will be continued under the 5th Framework Programme for the next three years.

As a result of the internationalisation of economies, the increased world-wide exchange of goods, populations and information, and the comparative weakening of nation-states within a supranational complex, European cities now seem to take the lead in the development of the continent. Although their political and institutional importance has yet to be fully recognised, these changes have undeniably altered their status. Where once they were a mere relay station for state powers, they have become a key element in the structuring and organisation of the European territory.

The Civitas proposal was to compare the changes occurring in European cities and to analyse how each city reformulates policies and ways of governing in response to these transformations. The aim was to analyse the new forms of urban governance by using three points of view. The first is by taking a spatial approach to the transformation of cities. The second is by identifying the changes in the economic structure of cities. Finally urban governance is understood by analysing social and urban fragmentation. Each of these three analytical perspectives was entrusted to a working group. They were expected to show how these changes called for a new formulation of the policies and a new design for urban government.

The aim was to answer the questions raised at the outset of the Civitas initiative. Let us take up some of those initial questions here.

If European reality is essentially urban, what is the meaning of the urban notion today? Which new changes have occurred in the urban structures during the last 15-20 years – that is after the intensive urbanisation of the post-war period? Which new ways of organising urban space have developed? Which changes have taken place in urban life? What are the new roles of cities in the economic globalisation process? How can cities manage both economic competition and social cohesion? Which impacts do these changes have on the social structures? Is there an increasing social and spatial fragmentation of cities? Which effects do these developments have on the management of cities and urban policies? How do these various changes exacerbate the welfare and fiscal crises that affect the government of cities? Which new forms of urban governance have appeared?

To answer these questions, three working groups were established at the beginning of the Civitas initiative. In the course of the action, more than fifty researchers were involved in these working groups. They came from forty universities and institutions and represented various disciplines: geography, sociology, economy, political science, urban planning, and architecture. In many countries, participation was open to young researchers or post-graduates. Fourteen countries were members of the action: Austria, Belgium, Denmark, Finland, France, Germany, Ireland, Italy, The Netherlands, Norway, Slovenia, Spain, Sweden, and Switzerland. The books that we have published bring together the papers produced by the working groups during this period of time.

The first volume is entitled *Exchange and Stability in Urban Europe: Form, Quality and Governance*. Its objective is to explain the transformations in the spatial organisation of the European cities that have taken place over the past two decades. It tries to show in what ways these transformations force us to reconsider the traditional definitions of the city. It reviews the changes, starting from the processes of spatial fragmentation, the complexification of the systems of governance, and the transformation of relations of the inhabitants to public space. The book highlights the diversity of the urban forms in relation to the changes which affect the cities. On that basis, this book tackles the question of the quality of life in new urban spaces. It goes on to discuss the capacity of urban policies to evoke real changes in the city and to regenerate the systems of urban governance.

The second volume is entitled *Governing Cities on the Move: Functional and Management perspectives on Transformations of European Urban Infrastructures*. It explores the transformation of European cities and their way of governing themselves. The transformation is related to the question of urban infrastructure, giving particular attention to transport and communication systems. Using many examples from European cities, this book focuses on the structuring role of these urban infrastructures on the social, economic, and ecological performance of the cities in the long run. It underlines the fact that these infrastructures support strategies, both from a functional point of view (the question of urban accessibility) and from the perspective of managing and governing urban spaces.

The third volume, *Governing European Cities: Social Fragmentation, Social Exclusion and Urban Governance*, tackles a major problem which can affect the basis of the European democracies in the future. The mechanisms of social exclusion, social marginalization, social fragmentation or, simply, social differentiation, are particularly evident in our modern European metropolises. These mechanisms are likely to weaken the fundamental principle of "being together" which characterises European cities. First, the book clarifies some concepts, which make it possible to explore the fragmentation of the urban societies and to identify the diverse forms that it can take throughout Europe, today and in the future. Then this book treats some strategies that have been drawn up to ensure the regulation of these phenomena and to come to grips with their consequences in terms of urban governance.

The books appear at a point in time when European cities, more than ever, are at the heart of the dynamics of the continent. In fact, they may be right in the path of impending storms. We can only hope that this work will be a starting point for new research programmes on the European urban world.

In conclusion, I would like to thank the members of the Technical Committee Social Science of COST, its successive chairmen, Bo OHNGREN and Dirk JAEGER, and the successive scientific officers Michel CHAPUIS, Henrik GRAF, Gudrun MAASS, Vesa-Mati LAHTI for their help during this programme.

Of course, I would also like to thank the following researchers, who were involved in the action A9 Civitas. I hope I have not forgotten any of them.

Hans Thor ANDERSEN (Denmark), Harri ANDERSON (Finland), Roger ANDERSSON (Sweden), Ann-Cathrin ÅQUIST (Sweden), Antoine BAILLY (Switzerland), Michael BANNON (Ireland), Ludger BASTEN (Germany), Robert BEAUREGARD (USA), Hubert BEGUIN (Belgium), Maurice BLANC (France), Kerstin BODSTROM (Sweden), Mario BOFFI (Italy), Ivar BREVIK (Norway), Jack BURGERS (Netherlands), Anne COMPAGNON (Switzerland), Pasquale COPPOLA (Italy), Barbara CZARNIAWSKA (Sweden), Giuseppe DEMATTEIS (Italy), Frans DIELEMAN (Netherlands), Martin DIJST (Netherlands), Ingemar ELANDER (Sweden), Karl-Otto ELLEFSEN (Norway), Heinz FASSMANN (Austria), Jürgen FRIEDRICHS (Germany), Alex FUBINI (Italy), Oscar W. GABRIEL (Germany), Francis GODARD (France), Francesca GOVERNA (Italy), Vincenzo GUARRASI (Italy), Peter GÜLLER (Switzerland), Knut HALVORSEN (Norway), Brita HERMELIN (Sweden), Wolfgang JAGODZINSKI (Germany), Hubert JAYET (France), Bernward JOERGES (Germany), Gertrud JORGENSEN (Denmark), Dominique JOYE (Switzerland), Pavlos KANAROGLOU (Greece), Hans KRISTENSEN (Denmark), Klaus KUNZMANN (Germany), Isabella LAMI (Italy), Jolanda LESNIK (Slovenia), Jacques LEVY (France), John LOGAN (USA), Lienhard LÖTSCHER (Germany), Guido MARTINOTTI (Italy), Erkki MENNOLA (Finland), Markus MEURY (Switzerland), Sako MUSTERD (Netherlands), John NOUSIAINEN (Denmark), Giampaolo NUVOLATI (Italy), Willem OSTENDORF (Netherlands), Ursula REEGER (Austria), Paolo RIGANTI (Italy), Walter SCHENKEL (Switzerland), Harry SCHULMAN (Finland), Daren SCOTT (United Kingdom), Ann SKOVBRO (Denmark), Veronique STEIN (Switzerland), Jacques-François THISSE (Belgium), Isabelle THOMAS (Belgium), Ronald VAN KEMPEN (Netherlands), Ann VERHETSEL (Belgium), Jan VRANKEN (Belgium), Andres WALLISER MARTíNEZ (Spain), Ari YLONEN (Finland).

Claude Jacquier

Chairman of the Management Committee of the COST Action A9
"Civitas – Transformations of European Cities and Urban
Governance"

Acknowledgements

This book is the result of many challenging discussions between engaged scientists from nine countries in Western Europe. The cross-national network has opened up new angles and comparisons on similar problems in parallel circumstances. The experiences from our work clearly support the potential in multi-national collaboration, as such an approach allows an identification of institutional peculiarities and their importance for finding solutions.

In the last few months (October 1999 –September 2000) we have had intensive correspondence with each of the contributors. Many pre-final versions went between us, as editors, and the authors of the different chapters. However, we have not tampered with any of the chapters contrary to their author's wishes, and each remains the responsibility of their own author.

We would like to thank those who lent us a hand in these months. The English was corrected by Nancy Smyth van Weesep who, to our opinion, did a great job. All the corrections were typed by Ciëlle van Dooren, Franc Faay, and Marlijn van der Hoeven. They also checked the reference lists with the texts and made the final lay-out. Martin Dijst would like to thank the Urban Research Centre Utrecht (URU) of Utrecht University for the time they gave him for the work on this volume.

Martin Dijst, Utrecht
Walter Schenkel, Zürich
Isabelle Thomas, Louvain-la-Neuve

1 Urban Performance in Perspective

MARTIN DIJST AND WALTER SCHENKEL

Transformation of Cities

Competitiveness and sustainability are the keys to the long term future of the European Union's economy, creation of wealth and employment opportunities, enhancement of the quality of life of Europe's citizens, and preservation of the environment and the natural resource base (European Commission, 1997). These keys to Europe's future can be found particularly in cities and their agglomerations. By the turn of the century, almost half of the world's population will live in cities (WCED, 1987). At this moment, 80% of all Europeans have their homes in cities (Cavallier, 1998).

Urban agglomerations are focal points in the economic, social and cultural development of a region. These qualities are mainly the consequences of site and situation characteristics such as the size, density and diversity of their populations, economies of scale, synergies, valuable cultural and spatial characteristics inherited from the past, transport systems, etc. Some essential features of cities are the *large concentration* of specialised functions and their associated activities in a restricted area as well as the diversity of social classes (Kreukels, 1993; Krugman, 1991; Nijkamp and Perrels, 1994; Lambooy, 1998; Thisse and Van Ypersele, 1998). The post-industrial city is associated with the growth in services and, more recently, with the surge in communication technologies (Bairoch, 1985; Castells, 1991).

In the last decades, several processes have limited the growth of the cities and reduced their importance as sources of wealth (Button

1

and Pierce, 1989; Stanners and Bourdeau, 1995). Cities are now confronted with problems such as economic restructuring, changes in the composition of the population (e.g. foreign immigration, growth of the elderly population, growing income disparities), increasing mobility and consumption, congestion, pollution, poverty, crime, and/or a decline in ecological conditions. These developments show the vulnerability of cities to processes taking place in society at large (Storper, 1997). Three issues, which are strongly related to the competitiveness and sustainability of cities, need to be addressed by local management: segregation, economic performance and ecological problems.

Between cities and urban regions, *segregation processes* increase. Although there are some exceptions, like the French cities, suburbanisation of the wealthier population categories left most city centres with a preponderance of lower-income groups. Furthermore, fragmentation challenged the social cohesion within cities. Fragmentation may be defined as a specific state of social differentiation, namely a reduction of social ties among social groups, extremely low mobility between groups, and a great variation (disparity) of behavioural options (Friedrichs and Vranken, 2000).

Migration processes influence not only the social performance of cities but also their *economic performance*. These spatial processes are powered by the globalisation of the economy, the replacement of industry by the service sector, the widespread use of information and communications technologies (ICT), and acceptance of the car as the main transport mode. During the last decades, not only households but also facilities and firms moved out of many European cities. Some of them found a new place of business in the suburbs. As a consequence, an urban field has developed in the last decades (Lambooy, 1998). This massive movement meant an expansion of the functional urban territory; the urban system has still kept or even reinforced its original function (Nijkamp and Perrels, 1994). Although the urban system has expanded, the daily life of households as well as enterprises is for a large part limited to the urban region (Lambooy, 1998). Wiewel and Persky (1994) call this 'the growing localness of global cities'.

Because of a new spatial distribution of population and investments in new infrastructure for the private car, a new trade-off between production and transportation costs favoured locations outside the cities. As a consequence of these spatial processes, the dominant relations in the mobility patterns changed from radial into tangential. At the same time the private car became the most dominant transport mode (Gordon, et al., 1991; Schmitz, 1993). Hence, these developments increased the *ecological problems* of the cities. Congestion within European cities seems to hinder urban growth. However, urban transportation is a public domain in which policy has not been very effective. This often derives from a bad distribution of the responsibilities between the many parties involved. Intra-urban mobility is essential for the development of contacts that lead to the efficiency of the city. It influences the existence and the level of land rent. It models the development of the urban structure in the manner of a putty-clay technology (Thisse and van Ypersele, 1998).

These urban transformation processes are characterised by a growing diversity, complexity and dynamics, causing uncertainty among those who have to manage and control the urban transformation processes in order to improve the urban performance in economic, social and ecological respects. This uncertainty is increased in an era in which *the type of governance/management* changed. In the nineties, national governments in different European countries changed their relation with local governments. They reduced the financial flows from national to lower administrative levels. At the same time, they deconcentrated some former national tasks and gave these lower levels more freedom in performing these tasks. Besides, the relationship between the people (the governed) and the authorities (the governance) was changed (Cavallier, 1998).

Urban infrastructures are important to determine the effectiveness with which cities can compete economically and socially, all in the context of achieving sustainability (Convery, 1998). In order to improve this performance, urban managers need more insight in the effectiveness of land use, infrastructural policies and time policies. *Infrastructures* in the broadest sense influence the potential for urban development. Infrastructure is the basic (in Latin 'infra' means 'under') equipment which has been provided by human endeavour and which underpins the economic and social life of a community (Convery, 1998). Several types of infrastructure can be distinguished, among others:

- *communication and transport infrastructure*: public transport, motorways, parking places, harbours, teleports, etc.;
- *cultural and recreational infrastructure*: public places like squares, parks, playing fields, museums, and churches;
- *environmental services*: e.g. power supply, water supply and waste water treatment;
- *health services*: hospitals, dentists, family doctors.

This book is mainly confined to the first two types of infrastructure.

The investments in these urban infrastructures have a relatively long life. They are the concrete assets of cities; their provision and maintenance is a core preoccupation for policy-makers (Mega, 1998). Cities and their infrastructures are not made of LEGO, so the users have to cope with these 'historical' structures in their daily lives (Gordon and Richardson, 1989). The way in which local governments bring into action these urban infrastructures determines in great measure the social, economic and ecological performance of their cities for a long time.

A successful strategy for investment in infrastructures has to be developed from two perspectives:

- *a functional perspective*: We have to know in what way and to what degree the performance of individuals, companies and institutions will be influenced by the characteristics of the urban infrastructures and in what manner their behaviour will influence the performance of the cities which they use. Especially the way in which transportation systems, housing, companies and facilities are accessible is very important in this respect. Transportation and spatial policies influence accessibility, but so do time policies.
- *a governance/management perspective*: We have to know which type of governance/management is best suited to influence urban development, how different groups are connected to each other, and how actor networks are involved in political and

administrative decision-making processes. Process-oriented management variables are very dynamic.

The main objective of this book is to elaborate, in both theoretical and empirical ways, the functional and governance/management perspective with respect to investments in urban infrastructures. Those actors who want to improve the performance of urban areas by putting in infrastructures need to pay attention to both perspectives. The success of any investment strategy is dependent on how people as members of households, companies or institutions will use urban infrastructures in their daily lives and how decisions on investments are taken by the actors. Insights in these behaviours can help public and private actors to cope with diversity, complexity and uncertainty in a dynamic urban environment.

Functional Perspective

At the turn of the millennium, some economic and socio-cultural megatrends can be observed. The world economy is highly dynamic. The Information Revolution is sweeping across the world. The processing and the distribution of information with the help of ICT have become the main strategic activity in the global economy, determining the competitiveness of the national economies. At the same time, pushed by information technology, the economy is internationalising. National economies become dependent on other economies. Furthermore, the way the production and management of companies are organised has changed. Vertically organised large corporations are being replaced by flexible, horizontal network organisations (Castells, 1991).

From a socio-cultural perspective, we see that for decades, lifestyles in Western societies have been becoming less and less uniform. The time when most people spent their days in or around their home is long gone. Rising levels of affluence, the demise of traditional values, and the increasing impact of self-actualisation have led to a wider range of choice for individuals and households. As a result, more and more households combine tasks such as paid work, housekeeping, and perhaps childcare. This is illustrated by women's labour market participation. In 1995 in the European

Community, Sweden and Finland led the way: the ratio between women's and men's labour participation was almost one. On the other hand, in the Netherlands this ratio was approximately 0.7. Countries like Denmark, France, the United Kingdom, Germany, Austria and Portugal take a position in the middle (Dijst, 1999b).

In the wake of these developments, social, economic and cultural structures are increasingly being determined by persons, companies and organisations that maintain relations with activity places located elsewhere. In a *network society* such as this, the significance of physical distance declines and the importance of available time increases. Accessibility outweighs proximity. Drewe (1996) posits that for households in a network society, the spatial planning perspective on the accessibility of a particular activity place shifts to the accessibility of multiple activity places.

The time dimension of activity patterns is especially important. Time is not only a limited resource for participating in activities, but also a medium: 'It may be treated as a path which orders events as a sequence which separates cause from effect, which synchronises and integrates' (Cullen, 1978). As a consequence of the growing complexity of society, there is a tendency for people and companies to differentiate with respect to the way they are using time (see also the contribution by Boffi and Nuvolati in this book). This leads to an increasingly rich diversity of mobility patterns among individuals, households and companies.

Of course, it is not only the functioning of people and companies in their urban environment that determines the performance of the city or urban region. Also the performance of districts within cities and their region is relevant in this respect. These places are increasingly affected by the growing complexity of activity and mobility patterns of individuals and their households as well as by the mobility patterns generated by companies. The activities in which people as member of a household or company participate bring them at certain locations to certain times. It is impossible to be in two places at once. These locations can be some activity places like shops, services or offices. These locations can also be some public places in the 'open air', like parks, streets, squares, stations, airports and other public spaces. As inhabitants and visitors of cities differ in their ability to reach locations and urban areas, these places can be characterised by different populations at different times of the day. Bonfiglioli (1997; 1998) calls these areas 'chronotypes'; Goodchild and Janelle (1984) use the concept of 'temporal specialisation'. As a consequence, the performance of activity places or public spaces located in these areas is very strongly dependent on the activity and mobility patterns of the visitors.

Public authorities are losing their grip on these technological, economic, social, cultural and spatial developments. Processes are no longer limited to administrative territories and can no longer be directed by national governments alone. Different types of policy networks, which consist of public and private actors, have to be developed (see section 'Governance and management perspective'). Furthermore, the planning concepts have to be changed. Considering the direction in which society has developed, generic spatial concepts like growth centres, urban nodes, compact cities and compact city regions no longer suffice. But many planners apparently do not get the message. Time and time again, they come up with policy goals such as compact urban development and the containment of suburban expansion which give too little credence to the big differences between persons, households and companies with regard to how they can and want to use space. Also cities and regions are showing more and more differences in the composition of their population and the range of businesses and services.

In a network society, in which besides a 'space of places' a 'space of flows' develops (Castells, 1991), the *concept of accessibility* is central to the functioning of individuals and their households, facilities, firms and urban spaces. This concept refers to the ability to visit activity places – like shops, work places, services, companies, public spaces, etc. – by using a particular transport system at an acceptable time or financial cost. It is known that accessibility summarises information on the location of households and firms in relation to the distribution of activity places and the transport system that connects them. Hence, accessibility is an important criterion used to measure spatial structure and urban performance (see Morris et al., 1979, or Bruinsma and Rietveld, 1998 for a review).

In order to increase the effectiveness of policies at the local and regional level, thereby to improve urban performance, planners must understand the (options for) behaviour of individual actors. Besides, the plans have to be formulated to meet specific local and regional needs (Dijst, 1999a). Not all planners are aware of the influence their policies could have on accessibility and on urban performance. For instance:

- *spatial policies* (e.g. locational policy) can influence the distribution of activity places and the characteristics (e.g. size, location and mix of functions) of public spaces;
- by applying *time policies*, they can change the opening hours of services and timetables of public transport;
- *infrastructural policies* can modify the structure of the transport network, the supply of public transport stops, and the supply of parking places, whereby the relative advantage of different transport modes can be changed;
- *pricing policies* can change the financial attractiveness of different transport modes.

These policies will be discussed in different contributions in this book.

Urban performance can be described from two *perspectives*: that of the individual or household; and that of the activity place, like shops, services and enterprises or urban area. From the *individual/household perspective*, it is important to know which set of activity places a person can select, given his place of origin, as destinations at acceptable (time) cost. From *the perspective of an activity place/urban area*, the ease with which an activity place may be reached by individuals from other locations at acceptable (time) cost will be described. If the activity places or urban areas that lie within reach of a person are not meeting their needs or if not enough people can visit these places, the performance of both entities – person/household and activity place/urban area – are not optimal. In the end, this can harm the urban performance of a city or urban agglomeration.

On the basis of these two perspectives on urban performance, the contributions in the functional stream of this book can be characterised (Figure 1.1). The paper by Verhetsel and the one by Boffi and Nuvolati were written from the perspective of individuals/ households. They show us, respectively, the effects of individual travel decisions on traffic streams in the Antwerp region and the time use and mobility patterns of people residing in the

Individual/household	Activity place/urban area
Boffi/Nuvolati	Riganti
Verhetsel	Stein

Figure 1.1 Two perspectives on urban performance used in the four 'functional' case studies in this book

Milan urban area. The other two contributions show more interest in the performance of urban areas. Riganti shows us the effect of Milan's and Turin's Crossrail link system on the land-use structures in both cities. Stein is interested in the use of public spaces in Geneva.

Governance and Management Perspective

Developments in urban performance take place in an era in which the type of governance has also changed. The history of urban policy in the 1980s and 1990s is marked by three characteristics: a continued spatial targeting, an emphasis on the private sector and a rather non-hierarchical role for local authorities, and the time limits envisaged for the programmes (Hill, 1994). We assume that traditional planning policies are no longer able to manage such a wide range of urban phenomena; there is a shift from physical and spatial development planning to more integrated planning, covering socio-economic, environmental, institutional and financial variables. This means that top-down, centralised and hierarchical management of public policies has to be revised and transformed into a more decentralised, reticular, and interactive process (e.g. through public-private partnership). As a *spatial consequence*, we can assume that urban governance separating public space and private property is in many cases no longer adequate for solving planning problems in highly complex urban areas.

The governance perspective of this book puts emphasis on questions such as which *type of governance* is best suited to urban development, how *different actor groups* are connected to each other, and how *actor networks* are involved in political and administrative decision-making. Governance involves strategic activities, whereas the management perspective concerns operational problems. Governance is a way to directly influence social processes, to extend boundaries, to continue interactions between network actors, to transform game-like interactions into trust and rules, and to give relevant networks a certain autonomy (Kickert et al., 1997). Although the case studies cover a broad field of urban governance, the book seeks to contribute to the development of *common criteria* which can help compare urban planning aspects in different cities and countries. The comparison between different traditions may shed light on the changing role of governance, the role of 'new' strategies, and the relation between network qualities and learning processes. It is the *task of political theory* to clarify the choices that people are making in terms of their beliefs and to clarify the contents of these belief systems themselves. It is therefore obvious that the aim of this book should be to answer the question *Who can learn from whom?* – even when the results are expected to be much more complicated than the research question appears. Here, the so-called *argumentative turn* probably offers the possibility to go beyond traditional research approaches (Fischer and Forrester, 1993; Scott, 1991). It is imperative to build an analytical framework in which criteria for choice of instruments, the basic belief system of actors, and arguments justifying support or opposition can be set in relation to each other.

What types of management and governance are necessary for cities if they are to become *efficient* and *effective* in terms of economic, social and environmental performance? In view of the great number of actors and interests, we wonder if it is possible and meaningful to look for new ways to initiate planning processes that give priority to integrative targets for living, working and recreation in developing urban areas. Once a development problem has taken centre stage, a *new model of management* requires government agencies that can bring about processes of change. This involves bridging the gap between long-envisioned aspirations and short-run opportunity (Strauch, 1996). Urban planning projects can be an opportunity to open a so-called *policy window*. Once advocates of a solution sense that there is some movement, they leap in to promote their ideas (or to oppose established strategies) (Kingdon, 1995). Besides these *dynamic dimensions* of new forms of co-operation and communication, *stable dimensions* are extremely important too: each form of co-operation and communication is a sub-system of a superior political-administrative system. Framework conditions such as existing international, national and regional structures or the degree of legal, financial and organisational interlacing have to be systematically analysed.

From a theoretical point of view, it is assumed that urban planning activities, decision-making and implementation can no longer be understood as hierarchically structured processes. The traditional (hierarchical) authority arrangement must be considered as a constellation of more or less loosely connected political, economic and social actors. It is therefore suggested to conduct explanations of dynamic research dimensions according to some elements of the so-called *policy network approach* (Kenis and Schneider, 1991; Van Waarden, 1992; Scott, 1991). At the practical level, the network approach is oriented towards conditions on which actors may agree to implement effective strategies and achieve consensual targets. Networks are seen as consisting of a variety of actors, conflicting interests, and highly dynamic features; as an arena to make choices among problem-oriented, combined sets of alternatives; as platforms built upon divided power, unclear information, and a wide variety of goals.

The application of the network approach can contribute to a better understanding of a) determinants shaping the *discourse* within urban planning networks, and b) preconditions and functions of *learning processes* occurring within policy cycles in networks that produce planning decisions. A 'best solution' no longer refers to the success of the most powerful actor or the most feasible solution but an accepted and institutionalised arrangement to create optimal win-win solutions. Policy networks represent new forms of collaboration between state and society. Furthermore, derived questions are focused on changing network dimensions such as the shift from 'capacity-using' to 'capacity-giving' *use of power*. In other words the key issues are: *argumentative discourse* and *network management* instead of political struggle, dispute and confrontation. Network management means bringing in new ideas, reflecting on the network's features in the light of research results and evaluations, stimulating discourse, and facilitating learning processes (Kickert et al., 1997; Klijn, 1994). Here co-operative action is not only aimed at resolving problems but also at changing the context of the network. Managing the context means taking other policy fields and other actors – sometimes even new conflicts – into consideration.

		Dynamic dimensions		
		Actors	Instruments	Process
Stable	Politics	1,2	1,3	1,2,3
dimensions	Space	1,3	1,2,3	1,2
	Society	2	2,3	2

1) = Güller and Schenkel, 2) = Halvorsen, 3) = Lami

Actors:	administration, authority, company, political party, property owners, social interest group, economic interest group, business association, etc.
Instruments:	command-and-control norms, (non-)financial incentives, information, contractual agreements
Process:	formulation, planning, consensus-seeking, decision-making, implementation, evaluation
Politics:	formal political procedures (parliaments, governments, democracy), local political constellation
Space:	area structures, public space, private property, traffic and transportation network, up-grading potential, 'external' development projects
Society:	structure of population, employers and employees, living and work conditions, cultural activities, general economic situation

Figure 1.2 Case-study focus on urban governance

A political *programme* is defined as a cluster consisting of various resources: law, money, organisation and at least one measure. A policy *instrument* is defined as the way in which actors can modify one or more elements of a policy in order to gain a desired result (e.g. prohibition, subsidies, taxation, contractual agreements, information) (Linder and Peters, 1990). The increasing popularity of voluntary agreements, for instance, should be seen as an expression of the state's shifting role in many policy fields (Hanf and Koppen, 1993). *Strategy* combines programmes, measures, resources and instruments willingly and formally towards a governmental core target (Jänicke, 1997). On the basis of different aspects of the stable and dynamic dimensions concerning new forms of co-operation and communication, the governance and management case studies in this book can be characterised (Figure 1.2).

In summary, the comparison between different projects and countries should elucidate the changing role of governmental authorities and the role of new strategies. Furthermore, the public service orientation must pursue equity and due process for all citizens, as well as good standards, information and access for current consumers (Hill, 1994).

Outline of the Book

The main objective of this book is to elaborate theoretically and empirically upon the functional and management perspectives with respect to investments in urban infrastructure. Therefore, a book about urban governance must address three fields of inquiry: a) urban space and the desirable way of life in it; b) the most efficient combinations of actors producing collective goods and services; and c) principles of collective urban action and co-operation between elected and non-elected representation. In order to improve our comparative approach, the book is structured along the lines of three main topics: relationships between functional urban phenomena; infrastructure as a tool of governance; and actor-oriented management of networks. We can now bring up some arguments concerning key theoretical concepts and issues. Subsequently, we can select case studies, choosing them from different urban contexts in different countries. The book is divided into three *main sections*:

I   a theoretical one, giving an overview of relevant concepts in the fields of functional and governance/management approaches;
II a core section, devoted to the case studies;
III a concluding section, evaluating and comparing the case studies in the light of relevant concepts.

I Transportation, land use and accessibility are at the centre of two *theoretical contributions*. Dijst, Jayet and Thomas develop a conceptual framework to study urban performance from a transportation perspective. A key concept of this framework is accessibility. The authors show the influence of accessibility on daily mobility and on the locational decisions of households and facilities. The controversial issue of the relationship between transportation and land use is explained by Kanaroglou and Scott. These authors give an introduction to integrated urban transportation and land-use models. Their role of such models in policy analysis is to capture the important relationship in the urban system, so that the consequences of alternative policy decisions can be projected and studied in advance.

Mennola's contribution describes historical developments and characteristics in the political and administrative framework of European cities. The problem of complexity and the change from national to multinational operators is analysed in a comparative perspective, presenting theoretical concepts and giving examples concerning Max Weber's ideal types of unitary, village, multi-level and private cities. He presents an alternative way to organise, to analyse and to publish single case studies by linking them in comparative terms with their urban political and administrative environment. Schenkel's contribution serves as the theoretical and methodological basis for the book's actor- and process-oriented case studies. Urban planning processes are not only seen as continuous steps from policy formulation to decision-making and implementation but also as complex policy networks and interlaced policy cycles. Derived questions are focused on changing network dimensions such as the shift from 'capacity-using' to 'capacity-giving' forms of power in urban policy-making and new forms of co-operative planning.

II The *case-study section* includes contributions, which emphasise functional aspects or related urban phenomena as well as the role of actor networks, process management, political institutions and actor behaviour. Boffi and Nuvolati assume that the growing complexity of the urban system generates a multi-layered timing based on the fragmentation and combination of labour, consumption and social activities. Therefore, they develop specific typologies of time uses and time policies, which are relevant for urban areas and urban life styles. For Milan, they analyse the mobility patterns of different population groups, which are using the city. Riganti and Lami both analyse the same development but from a different perspective, looking at the new Crossrail systems in Milan and Turin. Riganti describes the potential effects on land use and travel demand. Lami analyses the financial feasibility of the Italian Crossrail projects, looking in particular at the principal players and promoters and their motives. Verhetsel presents an interesting multi-modal model for commuter traffic in Antwerp. Stein explores the meaning of public space in inner-city up-grading projects in Geneva. Physical infrastructure, transportation systems and public space are seen as a lever for change, able to evoke major transformation while guiding the change in a general framework. The contribution by Halvorsen and the one by Güller and Schenkel investigate renewal processes of old industrial areas, respectively in the Akerselva River Basin to the north of Oslo and in the Zurich River Basin. Both papers follow a set of relatively stable variables – historical developments, problem attributes and political frameworks – and a set of more dynamic variables such as network structures and actor relationships. Finally, network governance and process management are assumed to cause policy changes, changes within networks, new forms of co-operation and learning processes.

III In the *third section*, the editors review all contributions and formulate an answer to the theoretical questions, based on the results of the case studies. Especially in the context of cross-national comparative analyses, one should look more closely at the cultural and traditional background shaping the decision to adopt one strategy or the other. It is obvious that the book should try to answer the question 'Who can learn from whom?' – even through the results are expected to be much more complicated than the research questions seem to suggest.

References

Bairoch, P. (1985), De Jericho a Mexico. Villes et economique dans l'histoire, Gallimard.

Bonfiglioli, S. (1997), Che cos'e un cronotopo, in S. Bonfiglioli and M. Mareggi (Eds.) *Il tempo e la città fra natura e storia: atlante di progetti sui tempi della città,*

Milano, Politecnico di Milano, Departimento di Scienze del Territorio (Urbanistica Quaderni 12), pp. 90-92.

Bonfiglioli, S., Brioschi, L., Mareggi, M. and Pacchie, C. (1998), *Monitoring Developments in Working Time Organisation: Italian National Report*, Politecnico di Miliano, Dep. Di Scienze del Territorio, Milan.

Button, K.J. and Pierce, D.W. (1989), Improving the Urban Environment: How to Adjust National and Local Government Policy for Sustainable Urban Growth, *Progress in Planning*, 32, pp. 135-184.

Castells, M. (1991), *The Informational City: A New Framework for Social Change*, Centre for Urban and Community Studies, University of Toronto (research paper 184), Toronto.

Cavallier, G. (1998), *Challenges for Urban Governance in the European Union*, European Foundation for the Improvement of Living and Working Conditions, Dublin.

Convery, J. (1998), *Challenges for Urban Infrastructure in the European Union,* European Foundation for the Improvement of Living and Working Conditions, Dublin.

Cullen, I.G. (1978), The Treatment of Time in the Explanation of Spatial Behaviour, in, T. Carlstein, D. Parkes and N. Thrift (Eds.) *Human Activity and Time Geography*, Edward Arnold, London, pp.27-38.

Dijst, M. (1999a), Action Space as Planning Concept in Spatial Planning, *The Netherlands Journal of Housing and the Built Environment*, 14, 2, pp. 163-182.

Dijst, M. (1999b), Two-earner Families and their Action Spaces: A Case Study of Two Dutch Communities, *Geojournal* 48, pp 195-206.

Drewe, P. (1996), *De netwerk-stad VROM: bijdrage van informatietechnologieën aan nieuwe concepten van ruimtelijke planning*, vakgroep Stedebouwkunde, Faculteit Bouwkunde, TU-Delft, Delft.

European Commission (1997), *Towards the 5th framework programme*, Brussels.

Fischer, F. and Forrester, J. (Eds) (1993), *The Argumentative Turn in Policy Analysis and Planning*, UCL Press, London.

Friedrichs, S. and Vranken, S. (2000), European urban governance in fragmented societies, in H.T. Andersen and R. van Kempen (Eds.) Governing European Cities: social fragmentation, social exclusion and governance (forthcoming).

Goodchild, M.F. and Janelle, D.G. (1984), The City around the Clock: Space-time Patterns of Urban Ecological Structure, *Environment and Planning A*, 16, pp. 807-820.

Gordon, P. and Richardson, H.W. (1989), Gasoline Consumption and Cities: A Reply, *Journal of the American Planners Association*, 55, pp. 342-345.

Gordon, P., Richardson, H.W. and Jun, M J. (1991), *The commuting paradox: evidence from the top twenty*, Journal of American Planning Association, 57, pp. 416-420.

Hanf, K. and Koppen, I. (1993), Alternative Decision-Making Techniques for Conflict Resolution. *Environmental Mediation in the Netherlands* (Research Programme 'Policy and Governance in Complex Networks', no. 8), Rotterdam.

Hill, D.M. (1994), Citizens and Cities. *Urban Policy in the 1990s*, Harvester Wheatsheaf, New York/London.

Jänicke, M. (1997), The Political System's Capacity for Environmental Policy, in M. Jänicke and H. Weidner (Eds.) *National Environmental Politics: A Comparative Study of Capacity Building*, Springer, Berlin, pp.1-24.

Kenis, P. and Schneider, V. (1991), Policy Networks and Policy Analysis: Scrutinizing a New Analytical Toolbox, in B. Marin and R. Mayntz (Eds.) *Policy Networks. Empirical Evidence and Theoretical Considerations*, Campus, Frankfurt a. M. and Boulder Col, pp. 25-62.

Kickert, W. J. M, Klijn, E-H. and Koppenjan, J.F.M. (Eds.) (1997), *Managing Complex Networks. Strategies for the Public Sector*. Sage, London/Thousand Oaks/New Delhi.

Kingdon, J. W. (1995), *Agendas, Alternatives, and Public Policies*, New York, Harper Collins.

Klijn, E-H. (Eds.) (1994), *Managing Networks in the Public Sector: a Theoretical Study of Management Strategies in Policy Networks*, Eburon, Rotterdam.

Kreukels, T. (1993), Stedelijke Nederland: de actuele positie vanuit sociaal-wetenschappelijk gezichtspunt, in J. Burgers, A. Kreukels and M. Mentzel (Eds.), *Stedelijk Nederland in de jaren negentig: sociaal-wetenschappelijke opstellen*, Utrecht, Jan van Arkel, pp.9-37.

Krugman, P. (1991), *Development, Geography and Economic Theory*, MIT Press, Cambridge (Mass.)

Lambooy, J.G. (1998), *Agglomeratievoordelen en ruimtelijke ontwikkeling: Steden in het tijdperk van de kenniseconomie*, Faculteit Ruimtelijke Wetenschappen, Universiteit Utrecht (oratie), Utrecht.

Linder, S.H. and Peters, B.G. (1990), The Logic of Public Policy Design: Linking Policy Actors and Plausible Instruments, in Stuart S. Nagel (Eds.), *Policy Theory and Policy Evaluation. Concepts, Knowledge, Causes, and Norms*, Greenwood Press, New York, pp. 103-119.

Mega, V. (1998), The European City, in G. Cavellier (Ed.) *Challenges for Urban Governance in the European Union*, European Foundation for the Improvement of Living and Working Conditions, Dublin, pp.1-10.

Morris J., Dumble P. and Wigan M. (1979), Accessibility Indicators for Transport Planning, *Transportation Research*, 13A, pp. 91-109.

Nijkamp, P. and Perrels, A. (1994), *Sustainable Cities in Europe: a Comparative Analysis of Urban Energy-Environmental Policies*, Earthscan Publications, London.

Schmitz, S. (1993), Tangential statt radial: langfristige Veränderungen der Berufspendlerströme in Deutschland, *BfLR-Mitteilungen*, februar, pp. 5-6.

Scott, J. (1991), *Social Network Analysis. A Handbook*, Sage, London/Newbury Park/New Delhi.

Stanners, D. and Bourdeau, P. (1995), *Europe's Environment: The Dobris Assessment*, European Environmental Agency, Copenhagen.

Storper, M. (1997), *The Regional World: Territorial Development in a Global World*, Guildford Press, New York.

Strauch, V. (1996), Zur Entwicklung des Stadtforums, in H. Kleger, A. Fiedler and H. Kuhle (Eds.) *Vom Stadtforum zum Forum der Stadt*, Fakultas, Berlin, pp. 85-104.

Thisse, J-F and Van Ypersele, T. (1998), *Localisation des Activites Economiques: Efficacite Versus Equite*, Treizieme Congres des Economistes Belges de Langue Francaise, Charleroi.

Van Waarden, F. (1992), Dimensions and Types of Policy Networks, *European Journal of Political Research*, 21, Special Issue: Policy Networks, pp. 29-52.

World Commission on Environment and Development (1987), *Our Common Future*,

Oxford University Press, Oxford.

Wiewel, W. and Persky, R. (1994), The Growing Localness of Global Cities, *Economic Geography*, 70, pp. 129-143.

2 Transportation and Urban Performance: Accessibility, Daily Mobility and Location of Households and Facilities

MARTIN DIJST, HUBERT JAYET AND ISABELLE THOMAS

Introduction

As elaborated in chapter 1, urban agglomerations are focal points in the economic, social and cultural development of a region. These agglomerations are *large concentrations* of specialised functions and their associated activities and have a large diversity of social classes (Kreukels, 1993; Krugman, 1991; Nijkamp and Perrels, 1994).

A determining factor of these large concentrations is the availability of a local transportation system. The history of most large cities has been driven by technological change in urban transportation systems (Bairoch, 1985; Duranton, 1998). Moreover, transportation infrastructures are considered as one of the main instruments in the toolbox of land-use planners (Haggett and Chorley, 1972; Taaffe et al., 1996). Many decision-makers interested in the role of transportation infrastructure take it for granted that more infrastructure is always better than less because it leads to less congestion and/or better accessibility to existing facilities.

Such an argument is probably correct in the short run. But it is fair to say that our understanding of the long-run implications of such a policy is rather limited. Many questions remain unanswered about the marginal effects of policy-induced changes in the existing transportation infrastructure on the pattern of land use or on the urban form. It is not clear, for example, whether adding to the road infrastructure reduces congestion and vehicle emissions, or if it leads to a more dispersed and inefficient pattern of land development. Hence, measuring the accessibility of the urban areas and the efficiency of the transportation network are two interesting methodological topics; they both warrant further research.

Moreover, chapter 1 has shown that urban planners now face processes which tend to threaten the performance of cities in a social, economic and ecological respect. These processes take place in an era in which governance is changing too. In the nineties, national governments in Europe changed their relations with local governments. Urban performance has to be improved to reduce cities' social, economic and ecological problems, which are a threat to society at large. We also need to stimulate the social, economic and cultural developments of cities on which the performance of society is dependent. At the same time, it is increasingly acknowledged that more infrastructure may have detrimental effects. Therefore, in order to improve a city's performance, urban managers need more insight in the effectiveness of land use and transport policies.

This chapter presents a conceptual framework for studying urban performance from a transportation perspective in effort to identify and define research issues and concepts related to urban performance. A key concept in this framework is accessibility, which is defined and analysed in section 'Infrastructure, Accessibility and Reach'. In general, accessibility refers to the ability to visit activity places (shops, work places, services etc.) by using a particular transport system at an acceptable cost in terms of time or money.

All other things being equal, locations with inadequate access to activity places could hinder the daily performance of the households and the business. Ultimately, poor performance at the individual level could harm urban performance. Differential accessibility among modes is also important. Households determine at what time and/or financial cost they will travel to activity places. If the cost of reaching relevant activity places by bicycle or public transport is beyond this acceptable level, they probably will take the car to participate in activities. The resulting prevalence of the passenger car may lower the city's overall performance because individuals do not take congestion and pollution effects into account. In order to change their choice, we need more insight into the determinants of the daily performance of households. This topic is discussed below in section 'Transport Systems and daily Mobility'.

We also need to know how accessibility influences locational decisions of households and facility managers. This question is analysed in sections 'Transportation Systems and Facility Locations' and 'Household Location and the Land Market'. These decisions will influence the activity places that are reachable by individuals on a daily level. The network structure of the transportation system creates nodal points where facilities tend to concentrate. Competition for more accessible places influences land values. Land values and the willingness of households and other agents to pay the price, in turn, determine who can locate where. The joint effects of all agents' locational choices are manifest in the city's structure and urban dynamics (see also chapter 1).

Some general conclusions are proposed in the concluding section. They concern future research on the relation between urban performance and accessibility.

Infrastructure, Accessibility and Reach

First of all, urban transportation systems influence urban efficiency by determining the level and intra-urban distribution of *accessibility*, defined as the ability to visit activity places by using the transport system at an acceptable cost in terms of time or money. This ability (accessibility, in a general sense) can be described from two perspectives: the perspective of the individual/ household or the perspective of the activity place/urban area (see Figure 2.1).

Here, we refer to the first perspective as *reach* and the second one as *accessibility in a narrow sense* (Dijst and Vidakovic, 1997). Reach denotes the space in which a set of activity places is located, places that a person can choose from his place of origin as destination at an acceptable cost (in time or money). Accessibility (in a more narrow sense) denotes the space in which a group of persons is located and who, from their place of origin, can choose the activity place as a destination at an acceptable cost.

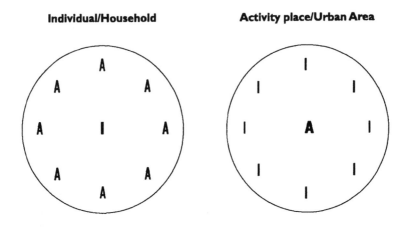

Figure 2.1 Two perspectives on accessibility:
the individual/household and activity place/urban area
perspective

Besides the two perspectives on accessibility, the measurements of accessibility are important. These have been developed gradually from early, partial and simple toward more complex and integral approaches. As the literature shows, it is not simple to quantify accessibility in a generalised form. In fact, the diverse nature of accessibility makes it difficult to employ a unique measurement scheme (Vickerman, 1974; Pirie, 1979; Jones, 1981; Lee and Lee, 1998; Bruinsma and Rietveld, 1998). Accessibility measures are numerous and can include the impedance effects of distance, time and generalised transport costs to produce a single index for each location (Linneker and Spence, 1992; Bruinsma and Rietveld, 1998). Accessibility research has led to quite a few papers, including studies of accessibility indicators (Shimbel 1953; Harris, 1954; Vickerman, 1974; Rich, 1980; Linneker and Spence, 1992), studies of the use of accessibility as an evaluation criterion for alternative transport plans (Spence and Linneker, 1994; Murayama, 1994), or studies of travel demand *models (Ben-Akiva and Lerman, 1975)*.[1]

The two basic elements of an accessibility measure are:

- information about the spatial friction affecting moves between places;
- information about place attraction, or the possibility which they offer.

The combination of both elements leads to the most usual family of accessibility measures, *gravity models*. Taking account of spatial friction leads first to *distance* measures; then it leads to *topology* measures which, instead of absolute distance, express the reach in terms of a number of connections (from one or more locations) offered by a network; and subsequently it leads *to cumulative opportunity* measures which indicate a number of places ('opportunities') that can be reached from one origin within certain distances or travel times (Black an Conroy, 1977; Breheney, 1978; Mitchell and Town, 1977; Stouffer, 1940).

According to Jones (Pirie, 1979), those indicators which only measure some characteristics of locations can be called *place accessibility measures*. Besides these, there is another category, which also accounts for characteristics of persons who are present at these locations. These are called *person accessibility measures* (Pirie, 1979). They are based on the fact that when a person leaves home, he or she generally visits not one but multiple activity places before returning (Pirie, 1979; Damm, 1979). Moreover, people take into account the amount of time available for travel. This depends, among other things, on the location of a future destination and the time at which one should be there (Burns, 1979; Dalvi, 1979).

Both reach and accessibility can be measured in terms of place and person accessibility. Which accessibility measure will be chosen for a particular investigation depends on the level of analysis and the availability of data. At an aggregate level, data are used to compare average travel characteristics of neighbourhoods, cities or other spatial units. At a disaggregate level, differences in travel behaviour between individuals in different spatial contexts are analysed.

Transport Systems and Daily Mobility

Society is becoming more and more complex. This shows up in an increasing diversity of activity and mobility patterns of individuals, households, companies and organisations (Dijst, 1999). The increasing differentiation in use of time and space between individuals and their households requires measures of 'reach', which explicitly account for time-spatial characteristics of population categories. The use of time and space is strongly conditioned by individuals' basic places (e.g. home and work), also called bases. These bases structure the activity and travel pattern of an individual. Cullen and Godson (1975) pointed out that: "Activities to which the individual is strongly committed and which are both space and time fixed tend to act as pegs around which the ordering of other activities is arranged and shuffled according to their flexibility ratings". As observed by Cullen and Godson, the time available for visits to other activity places is bounded by the departure from a base and arrival at the same or another base. The start and ending time and the duration characterise this available interval. With increasing interval length, the range of an individual's choices becomes wider. The most obvious argument for this is the increasing maximum distance reachable and the area within that distance (Lenntorp, 1976; Kitamura et al., 1981). For longer intervals, the individuals have greater choice in the use of time, number and type of stops, staying time and travel time.

In Figure 2.2 we see for each figure two spatial axes and one time axis. We can identify Hägerstrand's daily prism, which compromises a set of positions in space-time for which the probability of being included in the individual path is greater than zero. The projection of this prism onto space gives the potential action space, also called 'reach'. This is the area containing all activity places which are reachable, subject to a set of temporal and spatial conditions. This set of conditions includes: (I) the types and locations of activity bases; (II) the available time interval; (III) the travel speed; and (IV) the travel time ratio, i.e. the proportion of available time spent on travel.

The general form of action space is elliptical. When there are two bases, the four variables mentioned above delimit the area, which is

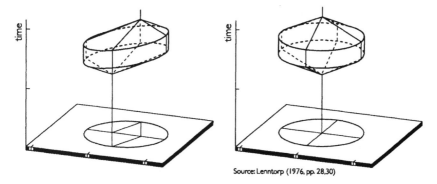

Source: Lenntorp (1976, pp. 28,30)

Figure 2.2   Reach or potential action spaces

reachable within the boundaries of an ellipse. Two other shapes of the action space are the line and the circle. When the whole time available has to be spent on the travel between the bases, the action space becomes a line: allowing no visits 'en route'. When the available interval starts and ends in the same base, the action space lies within a circle (Dijst and Vidakovic, 1997; Dijst, 1999).

In an era in which the ecological performance is very important for the development of cities, the choice between public and private transport, the latter being mainly the private car, is an issue of utmost importance. Even when households choose their transport mode on the basis of their lowest private cost, the overall transportation system may be inefficient. Households do not take account of the externalities they generate, the main ones being congestion and pollution, which becoming predominant in modern cities. Can we accept the ecological problems created by the prevalence of the private car? If not, how can we sell the idea of public transport to society?

The planning issues that urban managers are facing in order to change the detrimental mobility patterns are much more complex than ever before. For example, the provision of public transport that is tailored to a particular situation is a fairly novel concept. Everyone knows that a train or a bus does not take passengers to any corner of a city or city region at all hours of the day or night. If public transport is to remain affordable, some hard choices will have to be made. The timetable dictates when and where transportation will be available. Since the bus/train/tram does not have a stop at every corner of every street, it is necessary to provide feeder connections. Transportation must be tailored to the divergent demands of different people. The timetable and the location of the stops for trams/ trains/buses have to fit into the daily routines of the users. In view of the differentiation in household types that has appeared over the past decades, the planning task is now much more complex (Dijst, 1997).

Recently, an action space model called MASTIC (Model of Action Space in Time Intervals and Clusters) has been used in a Dutch new town, Zoetermeer.[2] The model is used to assess the opportunities of different household types to use sustainable modes of transport, such as mass transit and the bicycle. The model has four variables:

- distance between the bases of the action space;
- available time interval;
- travel time ratio;
- speed of travel.

The data are derived from fieldwork in which individuals were interviewed. On the grounds of that empirical research, we learned which types of activity places are visited in given time intervals ('the activity programme'). MASTIC calculates whether or not a person can carry out a desired activity programme within a specified time-space context. If the answer is affirmative, it is then determined which modes of transport can be used. If the answer is not affirmative, changes will have to be made in the time-space context. For example, the travel times on public transport can be improved if the location of services can be adjusted (Dijst, 1999).

The results of this study show that almost 57% of all activity programmes can be carried out on foot, by bicycle or by public transport. Action spaces with the home and the fixed working place as bases offer the best opportunities in this respect. Figure 2.3 shows the differences between household types. People over 65 years old have the best opportunities to use the sustainable transport modes instead of their own car in their daily life in Zoetermeer. More than 90% of their activity programmes in circle action spaces can be

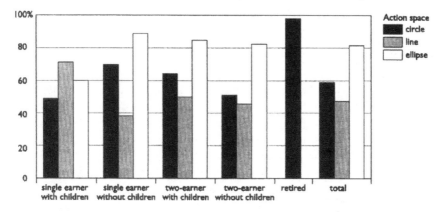

Source: Dijst et al. (1998)

Figure 2.3 Opportunities to use environmentally friendly transport modes for individuals from different types of households in Zoetermeer

carried out by using public transport or by biking or walking. Although two-income households are very pressed for time, for at least 50% of their activity programmes, they can use sustainable transport modes as an alternative for their automobile without losing 'much' time.

Transportation Systems and Facility Locations

Examining the impact of transportation systems on accessibility and reach is only a partial view. The transportation system does not determine accessibility alone. Moreover, in the long run, it does not influence accessibility and reach through the determination of transport time and costs only. Accessibility and reach are also determined by the intra-urban repartition of households and facilities, determining which facilities households can use within their action space. Since households compete for locations in accessible places, the following question arises: What is the impact of the network configuration on the intra-urban location of human activities and on urban development (if any)?

These facets of the problem have not been studied much.[3] The overwhelming majority of the contributions disregard the impact of (re) shaping the transport system to suit the locational pattern of human activities. This is, perhaps, because people involved in transportation analysis have (almost) no connections with those working in location theory. In this section, we examine this question from the point of view of facilities location. In the section 'Household Location and the Land Market', household locations will be examined from this perspective.

The literature devoted to *network location theory* generally deals with the problem of *where* to locate one or more facilities in order to achieve some objective function(s) under a set of constraints. Location-allocation models are concerned with the location of facilities to serve the distribution of clients best. Thus, models in this *locate* facilities and *allocate* individuals to them. Their interest is based on the commonly known equity-efficiency problem. This is obviously a very important family of problems with countless applications. Such location problems arise in many design tasks – where to locate facilities, plants, vehicles, people, services or any other system within a region or within a city. Nowadays, several useful operational research tools are available (Drezner, 1995; Labbé et al., 1995; Francis and Mirchandani, 1990). Facility location models have been developed to help the decision-maker in assessing the (social or private) benefit of different location systems. The geographical space is often represented by a graph, where the nodes are the demand points and/or the potential supply sites, and the edges represent the transportation network. Weights are assigned to the nodes (demand) and to the edges (transportation costs). In facility location analysis, the Hakimi theorem establishes that the search for a cost-minimising location along a network may be limited to the vertices of the network, thus showing that the facility location depends on where the nodes are (Handler and Mirchandani, 1979). These results are clear indications that the *shape of the transport network is likely to have a significant impact on the location of facilities*.

Optimal location problems often take place within a given transportation system. This system is often represented by a network, a graph with nodes (points in the discrete space representing communes, urban districts, etc.) and links connecting pairs of nodes (e.g. railways, waterways, roads). In this context, *transportation* depends upon the characteristics of the nodes (demand for travel) and occurs along the links; it is taken to connote the generalised costs of travel encountered by individuals in carrying out their activities or by firms in moving freight. By generalised costs, we mean some combination of monetary outlays, time length and/or efficacy of travel between specific locations. This way of considering transportation explicitly regards travel as generating negative utilities to the trip-maker. These prices are primarily a function of the supply of transportation infrastructure and of the demand for travel. The latter, in turn, is derived from the demand of individuals and firms for spatially distributed activities (e.g. employment, commercial outlets or residential locations) which generate and attract trips. Generically, these activities are referred to as *land-use activities*. The particular distribution and level of intensity of land-use activities are the key factors, which delineate the *spatial organisation* of regional and/or urban areas. The study of the interrelationships between land use and transportation has already been studied in *urban and regional economics* (Berechman et al., 1996). The literature often asserts that changes in the transportation system caused by – for instance – expansion of the road network will reduce travel time and costs. These effects, in turn, will encourage the dispersion of land-use activities, thereby altering existing patterns of travel demand and thus costs (Bonnafous, 1994). When do transportation costs decline? What does this mean to urban patterns of spatial organisation, i.e. the compact city versus the suburbanisation process and edge-cities developments? What does this new transport and communication system imply in terms of systems of cities? What does it imply in terms of regional or urban development?

The basic trade-off between fixed production costs and transportation costs lies at the heart of many location models (Beckmann and Thisse, 1986; Mulligan, 1984). That trade-off is often encountered in urban planning with respect to schools or recreational facilities, as well as in the design of a production-distribution-marketing strategy for a private firm (Erlenkotter, 1977). This trade-off is central to economic geography, where it appears in the pioneering analyses developed by Christaller (1933) and Lösch (1940). Indeed, the spatial configuration of human activities can be viewed as the outcome of a process involving centripetal as well as centrifugal forces. On the one hand, the existence of scale economies at the firm level is a critical factor for explaining the emergence of economic agglomerations. The mere existence of indivisibilities in human activities (Koopmans, 1957) makes it profitable for decision-makers to concentrate production in a relatively small number of facilities producing for dispersed consumers. Hence, increasing returns to scale constitute a strong centripetal force.

On the other hand, the need to interact among individuals and the corresponding transportation costs (defined broadly in order to include all impediments to mobility) imply that all activities are not concentrated in one place. In other words, the spatial dispersal of demand is a major centrifugal force. There is a fundamental *trade-off between scale economies and transportation costs* in the geographical organisation of human activities. As shown by Krugman (1991), this trade-off also underpins the organisation of the spatial economy at the multi-regional level. Depending on the relative strength of these two forces, a core-periphery structure might emerge as a stable outcome (Fujita and Thisse, 1996 for a detailed analysis). Consequently, it is fair to say that the trade-off between scale economies in production and transportation costs is critical for the geography of human activities. Thus, the trade-off will occur regardless of the particular institutional setting in which those activities are carried out. The urban environment could be a good example of further developments.

This conclusion must be qualified in view of recent contributions in spatial economics. As discussed by Arthur (1990) and Krugman (1991), human activities may also be locked in at some particular places for reasons that have nothing to do with the transportation network. Indeed, it seems that modern economies are more and more characterised by a putty-clay geography in which there is a priori a great deal of flexibility in the choice of locations but a strong rigidity in spatial structures once the process of agglomeration has started. The forces generating lock-in effects are based on the spatial interdependence between consumers and producers.

In order to gain more insight into the impact of transportation policy on the spatial pattern of facilities, Peeters and Thomas (1995) and Peeters, Thisse and Thomas (1998) consider different types of networks encountered in the real world. They study how the number and the locations of facilities are affected by the difference in the transport system by means of the simple plant location problem.

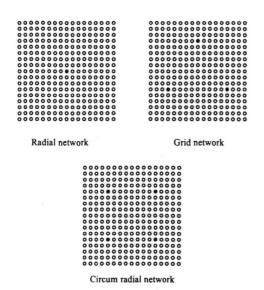

Radial network   Grid network

Circum radial network

Source: Peeters and Thomas (1995), Peeters, Thisse and Thomas (1998).

Figure 2.4 Optimal locations for the same modelling conditions but for three types of transportation networks

For example, they consider a squared lattice of points where each point is simultaneously a demand point and a potential location for a facility. The points are regularly and evenly spread over space (see Figure 2.4). Several transportation networks are designed on the same lattice of points: a *grid* network, a *radial* network and circumradial networks. In the last type of network, accessibility is improved because there are more edges. Since we focus on the impact of the transportation network, we assume that fixed costs are equal across locations in order to control for the role of differential factor endowments. Specifically, transportation costs are linear in distance while marginal production costs are zero.

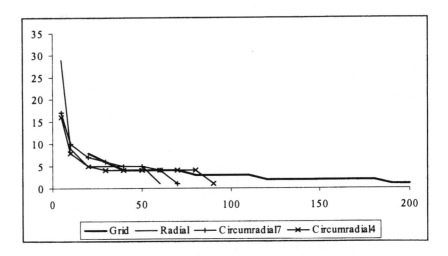

Figure 2.5 Example of location-allocation (UFLP model) output: variation in the number of facilities with fixed costs F and type of network (Circumradial7: circumradial with an external ring road; Circumradial4: circumradial with a central ring road)

Figure 2.5 shows that the optimal configuration of facilities contains less and fewer facilities as fixed costs rise. When the fixed costs are high, activities tend to concentrate in fewer places. In the radial network case, we see that the decrease is very sharp; a single facility solution is obtained from a fairly low value of the fixed cost, thus suggesting that the centre of a radial network is the source of a strong agglomeration force. On the contrary, in the case of the grid network, the optimal configuration involving a unique facility is obtained with a value for the fixed costs which is three times as large as in the radial case, confirming the impression that such a network yields more dispersal of human activities. For circumradial networks, we obtain intermediate solutions. This means that the construction of a peripheral road is indeed an instrument that can be used by spatial planners for the sake of fostering a more scattered distribution of human activities.

Let us now turn to the problem of characterising the optimal locations. Starting with a *radial* network, we observe immediately that a single facility is set up at the centre. This impression is reinforced by the fact that there is always a facility at the centre for all admissible values of fixed costs. On the other hand, a *grid* network leads to the construction of three facilities, which are evenly spaced. Interestingly, if a *peripheral* road is installed at some intermediate distance from the centre, the optimal configuration then involves four facilities, all located at the crossing of the radial roads and of the ring road. This shows that the attractiveness of the centre may completely vanish when a peripheral road is built at a well-chosen distance from the centre. It is worth noting that the choice of that distance (noted r) is crucial for this result: in the simulation sets where the distance varies from 1 to 8, it can be shown that the centre no longer accommodates a facility when $2 \leq r \leq 4$, but it is included in the optimal configuration for the other values for the distance. For small distances, the ring road has almost no impact on the optimal pattern of locations because the nodes it generates are too close to the centre. On the other hand, large distances are such that these nodes are now situated at the outskirts of the area. This means that they only have to supply small hinterlands, a fact that sharp reduces their attractiveness.

In short, Peeters, Thisse and Thomas demonstrate four points:

I   independent of the connectivity of the network, there is a relationship between the *shape of the network* and *the optimal spatial organisation of activities*, whatever the studied output of the model (locations, market areas, etc.);

II adding a *peripheral link* to a star-shaped network changes the location results. The location of the peripheral link is of great importance to the efficiency of the solutions: urban/regional economic growth depends strongly on a good location of the peripheral ring road. This raises the *centre-periphery problem* and has important implications for further empirical planning work. When location is of concern, the planner has to be aware of the importance of the location of a peripheral road;

III the *shape of the market areas* (zones) is very sensitive to the shape of the network;

IV transportation networks, especially radial networks, are to be viewed as a strong force of inertia in the location of human activities. *History* matters for locational patterns as far as transport networks are concerned. The formation of a common market is not likely to have a dramatic impact on the regional structures of locations, even though the absolute level of activities within each region may well be significantly affected;

V these simulations are limited to regional economics and to toy networks. Extending their experiments to urban areas would also mean introducing the effects of congestion in their model. That would require them to introduce negative externalities and the ensuing problems. Further work is needed along these lines (optimal location of human activities, shape of the transportation network, accessibility and externalities). Such studies are important as far as *urban performance* is concerned (theoretical developments, conceptual work, case studies and the role of congestion).

Household Location and the Land Market

The transport infrastructure and, more generally, the network configuration influences the intra-urban location of human activities and urban development through land and real estate markets. Within urban space, land is a scarce good. People ('agents') compete for land and, under normal competitive conditions, the agent buying or lending a parcel is the highest bidder. Therefore, the probability of an agent to be located on a specific parcel is proportional to the difference between the amount he is willing to pay and the amount other categories of agents are willing to pay. This bidding process is at the core of most recent urban economic models. (For an overview of these models, see Anas (1987); Fujita (1989); and Papageorgiou (1990)).

In turn, the knowledge of agents' willingness to pay, also called bid rent, helps us understand the current state and dynamics of intra-city locations. More importantly, this analysis leads to deeper understanding of the economic forces underlying spatial segregation processes in the city. Spatial segregation may be unintended: different social categories exert different demands for amenities, infrastructures and public goods and then choose different locations within the city. These differences appear in the analysis of bid-rent functions: *the larger the differences between bid-rent functions, the higher the unintended segregative forces.* (For an overview of unintended segregation mechanisms, see Fujita (1989) chapter 4.) Spatial segregation may also be intended, as when people try to choose locations occupied by the same category and far away from other categories. These processes appear in the determinants of the willingness to pay, which is influenced by the social mixture of the neighbourhood (Rose-Ackerman, 1977; Fujita, 1989).

How can we determine an agent's willingness to pay and its consequences on the bidding process? We must take into account the heterogeneity of real estate goods. Following Lancaster (1966), a house or a land parcel must be characterised by the whole set of attributes determining the utility level of its occupier (Arnott, 1985). Some of them are internal attributes (site characteristics) that describe the good itself: its size, composition (e.g. the number of rooms) and structure. The others are external attributes (situational characteristics) that describe the environment. Accessibility to infrastructure is one of them. But other factors also matter, for example the socio-economic mix of the neighbourhood. The price an agent accepts to pay for a specific good is a function of the whole set of attributes describing the good.

When households are homogeneous and able to move to the location where their utility level is the highest, real estate prices adjust to compensate for accessibility and amenity differentials.[4] Therefore, prices capitalise the benefit households receive from a more accessible location, any price differential being a monetary measure of this benefit (Alonso, 1964; Muth, 1969; Henderson, 1982; Fujita, 1989; and Papageorgiou, 1990). For urban planners, this information is important. "Capitalisation provides a natural measure of the social surplus, or willingness to pay for an increase in public goods. If this is so, and if a jurisdiction views land value as 'profit', which it tries to maximize, public goods should be provided efficiently" (Scotchmer and Thisse, 1995).[5]

If one wants to measure this capitalisation effect, one must know the whole price function, i.e. the function linking the characteristics of a real estate good to its price. A house price function of this type is estimated by Haughwout (1997), who uncovers evidence of a strong effect of a central city's infrastructure on housing prices in its surrounding suburbs. However, when households are heterogeneous, using the price function is not enough. As soon as the category of their occupier is not the same, the price differential between two homes no longer measures the willingness to pay of any category.[6] Therefore, when analysing land prices, one has to determine the shape of the willingness to pay (or bid rents) of the main categories of agents. The difficulty is that bid rents are not directly observed. They are implicit in the determination of real estate prices, the agent buying a home being, up to an arbitrary point, the highest bidder. Therefore, econometric analysis of land and house prices must combine features of the classical hedonic price (Rosen, 1974 and 1986) and generalised tobit models.

Conclusions

In the introduction to this contribution, we stated that urban agglomerations are focal points in the economic, social and cultural developments of a region or country. Several processes are threatening this valuable position of cities and their agglomerations. From a social and economic perspective, urban performance has to be improved in order to reduce the social, economic and ecological problems of cities and stimulate the positive sustainable developments mentioned earlier.

Urban performance at the aggregate level is directly related to the performance of households and firms at the individual level. From a transportation perspective, both kinds of performance are dependent upon the ability of individuals to visit activity places at an acceptable cost. If the activity places within reach of a person ('reach') do not meet one's needs or if not enough people can visit an activity place ('accessibility'), the performance of both person and activity places like facilities are not optimal. On the aggregate, this situation could hinder the performance of the whole city.

The time-spatial context of urban agglomerations (transport system and time spatial structure) and choices of the individuals concerning their life style, residential location, workplace and day scheduling determine both reach and accessibility. An important characteristic of the transportation system is the shape of the network. Network shapes differ by the degree to which reach and accessibility characteristics of locations are not uniformly distributed over the network. Consequently, people located in different network nodes will differ with respect to the transportation costs they have to pay to visit activity places. Accordingly, the performance of activity places like shops or public facilities is dependent upon the location of the customers and the transportation costs they have to or are willing to pay.

Households can influence these transportation costs through their choice a of residential and workplace location and their main transport mode. They will make a trade-off between transportation costs and their travel needs. In the same way, firms searching for locations make a trade-off between transportation costs and production costs. The results of these trade-offs are reflected in the land values or bid rents.

In order to improve urban performance from a transportation perspective, future research should focus on three questions:

- how and to what degree are location decisions and the performance of households and firms influenced by the spatial configuration of the transportation systems?
- how can these location decisions influence the economic and social performance of the city?
- how can city governments use spatial, transportation and time policy to change the performance of households as well as urban performance?

The conceptual framework presented in this paper is the first step towards a deeper understanding of the complex relations between performance of households and firms, shapes of infrastructure networks, land values, accessibility and urban performance. The second step will be to elaborate the basic ideas and formulate research projects.

Notes

[1] Accessibility techniques have also simply been used in order to identify the best locations for major facilities such as schools and hospitals (e.g. Robertson, 1976). Many case studies are, however, restricted to regional examples (see e.g. Spence and Linneker, 1994; Gutiérrez and Urbano, 1996; Dupuy and Stransky, 1996); urban case studies are less numerous (Laporte et al., 1994).

2   This municipality of almost 100,000 inhabitants is not far from The Hague.
3   For noticeable exceptions, see Peeters and Thomas (1995); Arnold, Peeters and Thomas (1997); and Peeters, Thisse and Thomas (1998).
4   For example, let us consider two houses, the first one being close to the infrastructure while the second one is far away from it. The only difference between the two houses is the accessibility to the infrastructure. There is a price differential between the two houses. This price differential is exactly equal to the amount of money a household accepts to pay for moving from the less accessible house to the most accessible.
5   A corollary of this argument is the well-known Henry George's theorem (Stiglitz, 1977): under mild conditions, a socially optimal level of public goods production is reached when these goods are fully financed out of a land tax.
6   Let us come back to the example given in note 3 with two homes, one close to the infrastructure and the other far from it. The agent who occupies the first home does not accept to pay for the second one at the current price, which implies that her willingness to pay for being close to the infrastructure is higher than the price differential. Conversely, the agent who occupies the home located far away from the infrastructure has a willingness to pay for being close to the infrastructure that is lower than the price differential.

References

Alonso, W. (1964), *Location and Land Use: Toward a General Theory of Land Rent*, Harvard University Press, Cambridge (Massachusetts).

Anas, A. (1987), *Modelling in Urban and Regional Economics*, Harwood, New York.

Anderson, W., Kanaroglou, P. and Miller, E. (1996), Urban Form, Energy and the Environment: a Review of Issues, Evidence and Policy, *Urban Studies*, 33 (1), pp. 7-35.

Arnold, P., Peeters, D. and Thomas, I. (1997), Circumradial Networks and Location-allocation Results. Is There an Optimal Location of a Peripheral Ring Road? *Urban Systems*, 1-2-3, pp. 69-90.

Arnott, R. (1985), Economic Theory and Housing, in E. Mills (Ed.), *Handbook of Urban and Regional Economics*, Elsevier Science Publishers B.V., Amsterdam.

Arthur, W.B. (1990), 'Silicon Valley' Locational Clusters: When Do Increasing Returns Imply Monopoly? *Mathematical Social Sciences*, 19, pp. 235-251.

Bairoch, P. (1985), *De Jéricho à Mexico. Villes et économie dans l'histoire*, Gallimard, Paris.

Beckmann, M.J. and Thisse, J.-F. (1986), The Location of Production Activities, in P. Nijkamp (Ed.), *Handbook of Regional Economics*, Elsevier Science Publishers B.V., Amsterdam, pp. 21-95.

Ben-Akiva, M. and Lerman, S. (1975), *Forecasting Models in Transportation Planning*. A paper prepared for presentation at the Conf. Population Forecasting for Small Areas.

Berechman, J., Kohno, H., Button, K., and Nijkamp, P. (Eds.) (1996), *Transport and Land Use*. Modern Classics in Regional Science: 2., Edward Elgar Publishing Company, Cheltenham.

Black, J. and Conroy, M. (1977), Accessibility measures and the social evaluation of urban structure, *Environment and Planning*, 9A, pp. 1013-1031.

Bonnafous, A. (1994), Réseaux de transport, in J.-P. Auray, A. Bailly, P.-H. Derycke and J. M. Huriot (Eds.), *Encyclopédie d'economie spatiale*, Economica, Bibliothèque de Science Régionale, Paris, pp. 325-332.

Breheny, M.S. (1978), The measurement of spatial opportunity in strategic planning. *Regional studies*, A, 12, pp. 463-479.

Bruinsma, F. and Rietveld, P. (1998), The Accessibility of European Cities: Theoretical Framework and Comparison of Approaches, *Environment and Planning A*, 30, pp. 499-521.

Burns, L.D. (1979), *Transportation, Temporal, and Spatial Components of Accessibility*, Lexington Books, Lexington.

Christaller, W. (1933), *Die zentralen Orte in Süddeutschland*, Gustav Fischer Verlag, Jena.

Cullen, I. and Godson, V. (1975), Urban networks: the structure of activity patterns, *Progress in Planning*, 4, 1, pp. 1-96.

Dalvi, M.Q. (1979), Behavioural Modelling, Accessibility, Mobility and Need: Concepts and Measurement, in D.A. Hensher and P.R. Stopher (Eds.), *Behavioural Travel Modelling*, Croom Helm, London, pp. 639-653.

Damm, D. (1979), *Towards a Modal of Activity Scheduling Behavior*, Massachusetts Institute of Technology, Cambridge.

Dijst, M. (1997), Spatial policy and passenger transportation, *Netherlands Journal of Housing and the Built Environment*, 12, 1, pp. 91-112.

Dijst, M. (1999), Action space as planning concept in spatial planning, *Netherlands Journal of Housing and the Built Environment*, 14, 2, pp. 163-182.

Dijst, M. and Vidakovic, V. (1997), Individual Action Space in the City, in D. Ettema and H. Timmermans (Eds.), *Activity-based Approaches to Travel Analysis*, Pergamon, Oxford, pp. 117-134.

Dijst, M., de Jong, T., Ritsema van Eck, J. and Vidakovic, V. (1997), *MASTIC-2: Model of Action Space in Time Intervals and Clusters*, Urban Research centre Utrecht, Utrecht.

Dijst, M., de Jong, T., Maat, C. and Ritsema van Eck, J. (1998), *Woonlocaties vanuit mobiliteitsperspectief*, Nethur/DGVH, Utrecht/Den Haag.

Drezner, Z. (Ed.) (1995), *Facility Location: A Survey of Applications and Methods*, Springer Verlag, Heidelberg.

Dupuy, G. and Stransky, V. (1996), Cities and Highway Networks in Europe, *Journal of Transport Geography*, 4 (2), pp. 107-121.

Duranton, G. (1998), La nouvelle économie géographique : agglomération et dispersion, *Economie et prévision*, 131, pp. 1-24.

Erlenkotter, D. (1977), Facility Location with Price-sensitive Demands: Private, Public and Quasi-public, *Management Science*, 24, pp. 378-386.

Francis, R.L. and Mirchandani, P.B. (Eds.) (1990), *Discrete Location Theory*, J. Wiley, New York.

Fujita, M. (1989), *Urban Economic Theory: Land Use and City Size*, University Press, Cambridge.

Fujita, M. and J.-F. Thisse (1996), Economics of Agglomeration, *Journal of the Japanese and International Economies*, 10, pp. 339-378.

Gutiérrez, J. and Urbano, P. (1996), Accessibility in the European Union: The Impact of the Trans-European Road Network, *Journal of Transport Geography*, 4 (1), pp. 15-25.

Haggett, P. and Chorley, R. (1972), *Network Analysis in Geography*, Arnold, London.

Handler, G.Y. and Mirchandani, P. B. (1979), *Location on Networks*, MIT Press, Cambridge (Mass.).

Harris, C. (1954), The Market as a Factor in Location of Industry in the United States, *Annals of the Association of American Geographers*, 44, pp. 315-348.

Haughwout, A.F. (1997), Central City Infrastructure Investment and Suburban House Values, *Regional Science and Urban Economics*, 27, pp. 199-215.

Henderson, J.V. (1982), Evaluating Consumer Amenities and Interregional Welfare Differences, *Journal of Urban Economics*, 1, pp. 32-59.

Jones, S.R. (1981), *Accessibility Measures: A Literature Review* (TRRL Report 967) Berkshire: Transport and Road Research Laboratory.

Kitamura, R., Kosttyniuk, L. and Uyeno, M.J. (1981), Basic properties of urban time-space paths: empirical tests, *Transportation Research Record* 794, pp. 8-19.

Kreukels, T. (1993), Stedelijke Nederland: de actuele positie vanuit sociaal-wetenschappelijk gezichtspunt, in: J. Burgers, A. Kreukels and M. Mentzel (Eds.), *Stedelijk Nederland in de jaren negentig: sociaal-wetenschappelijke opstellen*, Jan van Arkel, Utrecht, pp. 9-37.

Koopmans, T.C. (1957), *Three Essays on the State of Economic Science*, McGraw-Hill, New York.

Krugman, P. (1991), *Geography and Trade*, MIT Press, Cambridge (Mass.).

Labbé, M., Peeters, D. and Thisse, J.-F. (1995), Location on Networks, in M. Ball, T. Magnanti, C. Monma and G. Nemhauser (Eds.), *Handbook of Operations Research and Management Science: Networks*, North-Holland-Elsevier, Amsterdam, pp. 551-624.

Lancaster, K.J. (1966), A New Approach to Consumer Theory, *Journal of Political Economy*, 74, pp. 132-156.

Laporte, Y., Mesa I. and Ortega F. (1994), Assessing Topological Configurations for Rapid Transit Networks. *Research Paper CRT-999*, Montréal (Canada).

Lee, K. and Lee, H.Y. (1998), A New Algorithm for Graph-theoretic Nodal Accessibility Measurement, *Geographical Analysis*, 30 (1), pp. 1-14.

Lenntorp, B. (1976), *Paths in Space-time Environment: A Time Geographic Study of Possibilities of Individuals*, The Royal University of Lund, Department of Geography, Lund.

Linneker, B. and Spence, N. (1992), An Accessibility Analysis of the Impact of the M25 London Orbital Motorway on Britain, *Regional Studies*, 26 (1), pp. 31-47.

Lösch, A. (1940), *Die räumliche Ordnung der Wirtschaft*. Jena: Gustav Fischer.

Mitchell C.G.B. and Town S.W. (1977), Accessibility of various groups to different activities, *TRRL supplementary repat 258*.

Mulligan, G. (1984), Agglomeration and Central Place Theory: A Review of the Literature, *International Regional Science Review*, 9, pp. 1-42.

Murayama, Y. (1994), The Impact of Railways on Accessibility in the Japanese Urban System, *Journal of Transport Geography*, 2 (2), pp. 87-100.

Muth, R. (1969), *Cities and Housing*, Chicago University Press, Chicago.

Nijkamp, P. and Perrels, A. (1994), *Sustainable cities in Europe: A Comparative Analysis of Urban Energy-environmental Policies*, Earthscan Publications, London.

Papageorgiou, Y.Y. (1990), *The Isolated City State*, Routledge, London.

Peeters, D., Thisse, J.-F., and Thomas, I. (1998), Transportation Networks and the Location of Human Activities, *Geographical Analysis*, 30 (4), pp. 355-371.

Peeters, D. and Thomas, I. (1995), The Effect of the Spatial Structure on the p-Median Results, *Transportation Science*, 29, pp. 366-373.

Pirie, G.H. (1979), Measuring Accessibility: A Review and Proposal, *Environment and Planning*, 11A, pp. 299-312.

Rich, D.C. (1980), *Potential Models in Human Geography*. Concepts and Techniques in Modern Geography, 26, University of East Anglia, Geo Abstracts, Norwich.

Robertson, I. (1976), Accessibility to Services in the Argyll District of Strathclyde - A Location Model, *Regional Studies*, 10, pp. 89-95.

Rose-Ackerman, S. (1977), The Political Economy of a Racist Housing Market, *Journal of Urban Economics*, 4, pp. 150-169.

Rosen, S. (1974), Hedonic Prices and Implicit Markets: Product Differentiation in Pure Competition, *Journal of Political Economy*.

Rosen, S. (1986), The Theory of Equalising Differences, in O.C. Ashenfelter and R. Layard, *Handbook of Labour Economics*. Elsevier, Amsterdam.

Scotchmer, S. and Thisse, J.F. (1995), *Space in the Theory of Value: An Outlook and New Perspective* (mimeo).

Shimbel, A. (1953), Structural Parameters of Communication Networks, *Bulletin of Mathematical Biophysics*, 15, pp. 501-507.

Spence, N. and Linneker, B. (1994), Evolution of the Motorway Network and Changing Levels of Accessibility in Great-Britain, *Journal of Transport Geography*, 2 (4), pp. 247-264.

Stiglitz, J. (1977), The Theory of Local Public Goods, in M.S. Feldstein and R.P. Inman (Eds.), *The Economics of Public Services*, Macmillan, London, pp. 273-334.

Stouffer, S.A. (1940), Intervening Opportunities: A Theory Relating Mobility and Distance, *American Sociological Review*, 5, pp. 845-867.

Taaffe, E. J., Gauthier, H. L. and O'Kelly, M.E. (1996), *Geography of Transportation*, Prentice Hall, Upper Saddle River.

Vickerman, R. (1974), Accessibility, Attraction and Potential: A Review of Some Concepts and their Use in Determining Mobility, *Environment and Planning A*, 6, pp. 675-691.

3 Integrated Urban Transportation and Land-use Models for Policy Analysis

PAVLOS KANAROGLOU AND DARREN SCOTT

Introduction

As discussed in chapter 1, urban morphology has become increasingly complex over time, particularly throughout the latter half of the twentieth century. High rates of urbanisation, accompanied by rapid suburbanisation, have transformed the monocentric city, which has been the dominant urban form for several millennia, into one that is polycentric. The continuous decentralisation of population and employment from the core to the periphery of cities is well documented in the literature under the labels of *counterurbanisation* and *exurbanisation*. This phenomenon, articulated differently by Garreau (1991), refers to the development of *edge cities*, meaning urban centres at the outer limits of metropolitan areas.

Within these polycentric cities, processes that relate to everyday life have become more complex. Increasing household affluence, car ownership and female labour force participation rates impact household locational decisions and the way people organise their lives in cities. Furthermore, the emergence of a global economy, fuelled by fast-paced advances in transportation and communications technology, has an impact on the economic fabric of cities. Locational needs of firms operating in today's economies are decisively different than what they were only two decades ago.

Planning within such a fast-changing and complex environment has become increasingly difficult. Several authors have promoted the city as a complex system with several interacting subsystems (Bourne, 1982). Inherent in this conceptualisation is the idea that changes in one subsystem can affect and be affected by other parts of the system in unanticipated ways through direct relationships and through feedback loops. Considerable empirical evidence from around the world has documented the relationship between the transportation and land-use subsystems of cities. However, transportation policy and planning, more often than not, are formulated and practised independently from land-use considerations.

In the 1990s, however, there are signs that the status quo is changing in favour of a more unified approach to urban policy and planning. One of the major factors contributing to this is the maturation of the environmental movement, requiring cities, particularly the transportation sector, to reduce fossil fuel consumption in order to reduce harmful environmental emissions. Furthermore, it has become clear that any reduction in emissions achieved by technological improvements in cars will not be able to meet the sustainability goals. There must also be reductions in the distance driven for passenger and freight transportation in urban areas.

Because of the complexity of the urban system, policy designed to achieve this goal can have a wide range of unforeseen or even undesirable indirect consequences. The role of models in policy analysis is to capture the important relationships in the urban system so that the consequences of alternative policy decisions can be projected and studied in advance. Models that simulate the urban system are known as *integrated urban transportation and land-use models*, which, for the purpose of this chapter, are referred to as *integrated urban models* or *IUMs*.

Such models have a history of about three decades. Developments in the 1990s, in terms of computer hardware and software, database availability and sophisticated modelling methods, have renewed interest in such models. The time is ripe for the development of IUMs that can be adopted widely by planning agencies as useful policy decision support tools. Recent interest in such models has prompted the compilation of at least three critical reviews of existing models (Bureau of Transport Economics, 1998; Southworth, 1995; Wegener, 1994). The intention of this chapter is not to repeat what is found in these reviews. Instead, the objective is to provide a balance between a discussion of existing models and developments that will shape the next generation of IUMs.

Following this introduction, the urban system is described. We then provide an overview of the general structure of models that are currently used in practice. The purpose of this overview is to highlight the direction that basic and applied research on IUMs should take. The next section focuses on developments in *activity analysis* as the most promising area of basic research for the development of a new generation of IUMs that will explicitly account for the behaviour of actors in the urban system. Following this, we offer some concluding remarks.

The Urban System and its Management

Articulating the concept of a city and the processes that take place in it is not a trivial undertaking. Three useful terms in this context are: *urban form*, *urban interaction* and *urban spatial structure*. Urban form may be defined as the spatial configuration of fixed elements within a metropolitan area. Urban interaction refers to the flows of people, goods and information among different locations in the city. Urban spatial structure is a comprehensive concept. According to Bourne (1982), it consists of three elements: urban form, urban interaction and a set of organising principles that define the relationship between the two. The key point here is that urban form exerts a profound influence on the transportation flows within the city, but does not determine them completely. One can envision, for example, two very different commuting patterns that might overlay the same urban form. The first is one in which people commute to places of work close to their residences, while in the second, people commute longer distances bypassing nearby employment districts for more distant ones. In the first case, workers locate close to their place of work, making efficient commuting the organising principle. In the second, some other principle, such as proximity to relatives or people of the same ethnic origin and status, determines residential location. In reality, within a complex urban system several organising principles work concurrently.

The urban system is also a dynamic entity, its spatial structure evolving over time. Responsible for this evolution is a host of factors, not least of which is the behaviour of the actors – households, firms and public institutions – that operate in the system. The locational decisions of these actors can affect urban form. For example, empirical evidence suggests that the locational decision of a household is associated with its car ownership decision and ultimately the number and length of trips a household takes within a week. Thus, the behaviour of the actors is intimately related to the organising principles that give rise to urban spatial structure.

There are also natural and man-made processes that can shape the urban environment within which actors operate. Demography is a natural process that can have a pervasive impact on the dynamics of the urban system. Through *natural increase* (births minus deaths), the age distribution of the population can vary significantly over time. Different age groups demand different types of housing and goods, and follow different locational patterns. *In-* and *out- (net) migration* is another example of a demographic process that can affect an urban system.

The national and especially the regional state of the economy can have an impact on the type, number, size and location of firms operating in a metropolitan area, thus affecting the location and level of employment. Interest rates, one of the instruments used by governments to regulate the pace of growth of their economies, can have an impact on the demand and supply of housing, affecting the urban system. However, regional and especially local governments can have a direct impact on urban form through policies, such as the implementation of zoning bylaws, alterations in property taxation and changes in the transportation infrastructure and land use.

The obvious question that arises is whether the impact of government policy on the urban system can be anticipated before its implementation. If such anticipation is possible then policy can be used as a management tool in steering the urban system towards a desirable direction. The prerequisite for such a goal, however, requires a good understanding of the behaviour of actors in the urban system while taking into account the natural processes that take place in it. A considerable body of empirical literature has generated a wealth of knowledge, only a small portion of which is used to inform urban planning practice. An important finding in this literature is that the behaviour of actors in the urban system is such that the transportation and land-use subsystems are interdependent. The way activities are organised over space has a lot to do with the level of transportation demand. Conversely, supply in transportation infrastructure and services affects how activities are organised in space. The circular nature of the impacts between transportation and land use has been captured effectively by Wegener (1995), as shown in Figure 3.1, who argues for the integration of land-use and transportation planning.

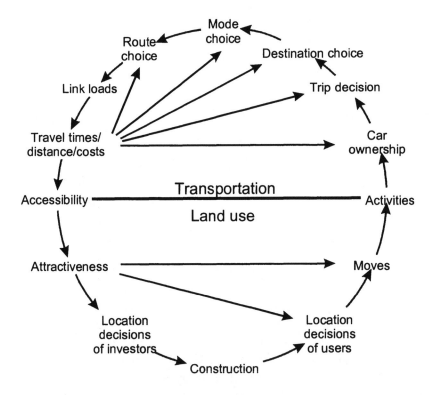

Source: Wegener (1995)

Figure 3.1 The land-use/transportation feedback cycle

In most developed countries, transportation planning is practised independently from land-use planning. Policy formulation and planning in the transportation sector, usually the domain of engineers and economists, falls under the jurisdiction of national governments. Land-use policy, on the other hand, practised by planners, architects and geographers, is the responsibility of local governments. Beyond the institutional, there are also ideological reasons for the existing practice. A parcel of land is usually owned and developed by private interests (households and firms) while the role of local authorities is to issue a permit. In contrast, transportation infrastructure is usually developed and owned by the state. The prevailing idea among transportation planners then is that land is developed in the open market and their role is to develop infrastructure that will accommodate the evolution of land development in the best possible way.

The 1990s have brought a change to this attitude, driven primarily by environmental concerns. The transportation sector has been identified as a major contributor of harmful environmental emissions in urban areas and many cities have set target emission reductions. The empirical literature suggests that improvements in automobile engine technologies will not be sufficient in meeting desired urban sustainability goals (Anderson et al., 1996a, 1996b). Reductions in the frequency and length of trips will be necessary. This realisation led to the conclusion that land-use policies that will bring the origins and destinations of motorised trips closer to each other must be designed and implemented. This idea has gained considerable institutional support. For example, the *European Commission's Green Paper* (CEC, 1990) is promoting compact cities as more sustainable than their sprawling counterparts.

In the United States, the *Intermodal Surface Transportation Efficiency Act* (ISTEA) became law in 1991. It transferred policy authority on transportation matters from the federal to local governments and required explicitly that transportation policy decisions consider the potential impacts on land use. The mandate of ISTEA was renewed in 1998 with the *Transportation Equity Act for the 21st Century* (TEA-21). A third piece of legislation, the *Clean Air Act Amendments* (CAAA) of 1990 also required a better co-ordination between transportation and land-use policy plans.

With these developments there has been renewed interest in integrated urban models as tools that could support decision making and could guide planners in the development, implementation and assessment of environmentally sound and fiscally responsible land-use and transportation policy. This is a proposition that poses considerable challenge to urban modellers. We now turn to examine the state-of-the-art and state-of-the-practice in urban modelling.

Integrated Urban Models: An Overview

The processes that characterise social systems are complex. For this reason, policy decisions can have long-term, unforeseen implications. The role of models in policy analysis is to assist in highlighting the undesirable consequences at the decision stage. Models, however, are not meant to replicate reality. They are meant to be a simplified picture of reality that captures the essential inter-relationships of the system under study, and as such, they are useful in informing policy decisions.

Models that are used in policy analysis can be *prescriptive* or *descriptive*. Prescriptive models, also known as normative, operations research or optimisation models, are used in situations where the objectives of a policy are known in advance. On the other hand, the purpose of descriptive models, sometimes called positive or conditional forecasting models, is to predict the possible outcomes of alternative policies. Sometimes descriptive models consist of a series of inter-linked sub-models, some of which may be prescriptive. Such is the case with existing IUMs, which makes their classification difficult. The mathematical structure for some of these models, however, is one optimisation framework, which puts them squarely into the prescriptive model class. This section discusses the ideas and concepts that have been incorporated into operational IUMs. We start by describing the *Urban Transportation Modelling System* (UTMS). Because the UTMS is still the preferred modelling tool for urban transportation planning by the vast majority of metropolitan areas, we refer to it in Figure 3.2 as the state-of-the-practice. We then discuss the basic ideas in the Lowry model, which provided the inspiration for the development of IUMs. Despite a history of about three decades of research on IUMs, a small proportion of metropolitan areas use them for transportation and land-use planning today, hence the reason for referring to them in Figure 3.2 as state-of-the-art. Many of the existing IUMs have borrowed ideas from urban economic theory and input-output analysis, which are discussed in turn.

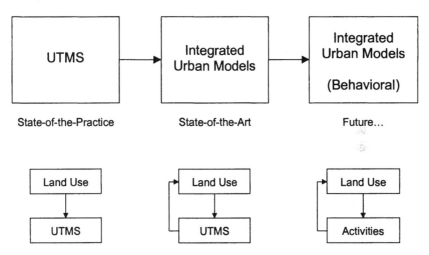

Figure 3.2 The evolution of integrated urban models

The State-of-the-Practice: UTMS

Urban transportation models are conceived as descriptive models. Early versions of such models appeared in the mid-1950s. By the 1960s, many metropolitan areas around the world adopted and used a variant of what is widely known as the Urban Transportation Modelling System (UTMS). At the bare minimum, the UTMS consists of four sub-models that are executed sequentially: *trip generation, trip distribution, modal split* and *traffic assignment*. The metropolitan area is represented with two layers of information, both of which are exogenous to the model. The first is a set of mutually exclusive zones. The second layer is a representation of the metropolitan transportation network. The distributions of land uses, population, employment and other economic activities for each of the zones are also given exogenously.

The trip generation sub-model uses regression models or categorical data analysis to determine the number of specific-purpose trips originating from each zone, as well as the number of trips destined to each zone (Institute of Traffic Engineers, 1987). The trip distribution sub-model accepts the information generated at the trip generation stage and produces a set of origin-destination trip matrices by trip purpose and time of day. The distribution of trips at this stage is achieved through the application of spatial interaction models (Fotheringham and O'Kelly, 1989; Wilson, 1970). More recently, logit models, based on utility maximisation theory, have also been adapted and used in the trip distribution stage as spatial interaction models. Trips that originate in a certain zone are allocated to other zones in proportion to the inverse of a function of impedance cost that separates the origin from its destinations.

The matrices generated at the trip distribution stage describe zone-to-zone person-trips. The purpose of the modal split sub-model is to obtain matrices of motorised trips, accepting as input the matrices generated at the trip distribution stage. The most popular model used for this purpose is the multinomial logit model, the parameters of which are estimated with survey type disaggregate observations on individual travellers (Ben-Akiva and Lerman, 1985; McFadden, 1973).

The purpose-specific, zone-to-zone motorised trip matrices that relate to the same time of day are added together. The resulting matrix serves as input to the traffic assignment sub-model. The objective at this stage is to identify the times needed to travel between zones in the metropolitan area, given that certain links of the transportation network are congested. Several algorithms have been developed for this purpose, including the *all-or-nothing* and *user equilibrium* algorithms (Sheffi, 1985).

The whole process starts with some initial estimates of zone-to-zone travel cost. After the first pass through all the sub-models a new set of travel costs is estimated. Then, the distribution, modal split and traffic assignment sub-models are run again with this new set of travel costs. The iterations are repeated until the travel costs stabilise.

The Lowry Model

It is probably a fair statement that all attempts to develop IUMs have been influenced by the seminal urban land-use model of Lowry (1964), which divides the economy of a city into two types of sectors: *basic* and *non-basic* (service). The basic sector, encompassing manufacturing and the primary industries, drives the local economy. The service industry exists to satisfy the demand for services of those working in the basic sector and their families. Thus, the size of the service sector is a function of the basic sector. The population associated with the basic sector is determined by multiplying basic employment by an exogenously provided city-wide ratio of total population to employment.

The Lowry model, just like the UTMS, assumes that the metropolitan area is divided into zones. There are three types of information that are exogenous to the model: the distribution of basic employment to zones, the land available for development in each zone and a matrix of transport costs among all the zones. Land use in each zone is determined through an iterative process. First, using the transport cost matrix, a spatial interaction model allocates the population related to basic employment to the zones of the city. This information is then used by a second spatial interaction model, which determines the location of service employment. The service employment for each zone is added to the basic employment and the process of allocating more service employment to the zones is repeated until stabilisation occurs. The model's output consists of the zonal distribution of residents and service employment.

The State-of-the-Art

ITLUP and other developments Several improvements to the Lowry model have been proposed over the years. Notable among them are those proposed by Echenique (1968) and Wilson (1974). The next major development as far as operational integrated urban models are concerned is the linkage of a land-use model, such as Lowry's, with a transportation model, such as the UTMS. The *Integrated Transportation and Land Use Package* (ITLUP) is credited for first implementing such a link (Putman, 1983, 1991). This model consists of linked sub-models, the *Employment Allocation Model* (EMPAL) and the *Disaggregate Residential Allocation Model* (DRAM), which are based on ideas taken from the Lowry model. Within these sub-models, trip generation and trip distribution are determined, thus accomplishing the first two stages of the UTMS. Three zone-to-zone matrices are produced, which provide the home-to-work, home-to-shop and work-to-shop trips. The sum of the three matrices is passed through modal split and traffic assignment models to determine congestion and estimate new costs (times) for travelling between zones. These costs are passed back to the DRAM/EMPAL combination and a new iteration starts. Iterations proceed until travel times stabilise. As Table 3.1 indicates, ITLUP has been used extensively – calibrated for more than 40 cities world-wide, most of which are in the United States.

Other existing operational IUMs have incorporated theoretical and methodological advances in urban and regional modelling. They could thus be classified according to the ideas that are incorporated in them. A first division of models into descriptive and prescriptive, however, is useful because it reflects differences not only on the foundation and conceptual basis of the models, but also on the purpose for their use. The formal structure of prescriptive IUMs consists of one constrained optimisation model, the structure of which usually turns out to be complex. The basic premise then of such an application is to assist planners and decision-makers in designing policies that will achieve system-wide efficiency.

The basic theoretical insight that provided the idea of developing prescriptive urban models was that spatial interaction models could be derived from *entropy maximisation*. Thus, a spatial interaction model can be written in an equivalent way as a mathematical program. Wilson et al. (1981) demonstrate how such a mathematical program can be embedded into a spatial activity allocation optimisation model, thus deriving a single mathematical program that will solve simultaneously the allocation of activities to zones and the interaction between zones. This and similar ideas have provided the inspiration for several prescriptive IUMs: BOYCE ET AL., KIM, POLIS, TRANSLOG and TOPAZ (Table 3.1). As the recent trend is to incorporate more behavioural realism into urban models, prescriptive mathematical programs can quickly become very complicated. Modelling the complex processes of an urban system within the boundaries of a single mathematical program leaves little room for expansion. There is room, however, for using such models within larger urban or regional modelling frameworks. Such is the case of POLIS, which is used within the Regional Economic-Demographic System by the Association of Bay Area Governments (Brady and McBride, 1992).

Table 3.1 Some integrated and empirically applied transportation and land-use models

Model	Useful references	Application
AMERSFOORT	Floor and de Jong (1981)	Amersfoort, Utrecht:Netherlands; Leeds: UK
BOYCE ET AL.	Boyce et al. (1992, 1993)	Chicago: USA
CALUTAS	Nakamura et al. (1983)	Tokyo, Nagoya, Okayama: Japan
CATLAS/NYSIM/ METROSIM	Anas (1983, 1992, 1995), Anas & Duann (1986)	Chicago, New York: USA
DORTMUND	Wegener (1982a, 1982b, 1986, 1995)	Dortmund: Germany
KIM	Kim (1989)	Chicago: USA
ITLUP	Putman (1983, 1991)	List includes San Francisco, Los Angeles, Houston, Dallas: USA
LILT	Mackett (1983, 1990a, 1991a, 1991b)	Leeds: UK; Dortmund: Germany; Tokyo: Japan
MASTER	Mackett (1990b, 1990c)	Leeds: UK
MEPLAN	Echenique et al. (1990), Hunt & Simmonds (1993), Hunt (1993, 1994)	List includes Bilbao: Spain; Sao Paolo: Brazil; Santiago: Chile; Naples: Italy
MUSSA	Martinez (1997a, 1997b), Martinez & Donoso (1995)	Santiago: Chile
OSAKA	Amano & Abe (1985)	Osaka: Japan
POLIS	Prastacos (1986a, 1986b), Caindec & Prastacos (1995)	San Francisco Bay Area: USA
PSCOG	Watterson (1993)	Puget Sound, Washington: USA
TRANSLOG	Boyce & Lundqvist (1987)	Stockholm: Sweden
TOPAZ	Brotchie et al. (1980), Dickey & Leiner (1983), Sharpe (1978, 1980, 1982)	List includes Melbourne, Darwin: Australia; Prince William Co., Virginia: USA
IMULATE	Anderson et al. (1994), Kanaroglou & Anderson (1997), Scott et al. (1997)	Hamilton: Canada
TRANUS	de la Barra (1989)	Caracas, La Victoria: Venezuela
URBANSIM	Waddell (1998a, 1998b)	Portland: USA

Source: Adapted from Southworth (1995)

Urban economic theory   Neo-classical urban economic theory, otherwise referred to as *bid-rent theory*, emanates from the work of Alonso (1964), and as recent work indicates, it has developed into a field in its own right (Papageorgiou and Pines, 1999). Households in the theory are conceptualised as trying to maximise their utility subject to a budget constraint. The utility is a functional form dependent on the consumption of land and a composite good, which makes up for all goods other than land consumed by a household. The budget constraint ensures that what households spend on transportation, land and the composite good does not exceed their income. The logical conclusion of the theory is that households locate in the city by trading off between transportation cost and what they pay for land. A spatial equilibrium emerges in which all households of the same income locate in the city in a way that they attain the same utility level. At equilibrium, no household can be better off by moving elsewhere in the city.

The basic idea behind this theory is that of equilibrium as a *market clearing mechanism*, which allows for the solution of land prices. The theory then predicts that the more accessible a location, the higher the price that it will be associated with. Early versions of the theory were developed for a monocentric city, where the CBD is the most accessible location, thus associated with the highest rent. As one moves away from the CBD, locations become gradually less accessible and the bid-rent declines monotonically.

Over the years, several suggestions have been made that would allow the incorporation of these ideas into operational simulation models. A fruitful one, by Anas (1984), was the development of a general equilibrium model for household/employment location and travel. This is derived from a combined utility and entropy maximisation problem. Ideas of this sort gave rise to descriptive IUMs (CATLAS/NYSIM/METROSIM, MUSSA, URBANSIM), as well as to prescriptive ones (KIM, BOYCE *ET AL.*).

Input-output analysis   Input-output (I-O) models, first introduced by Leontief in the 1930s (Leontief, 1941), constitute the primary tool for year-to-year economic forecasting by almost every national and regional government world-wide. Miller and Blair (1985) provide an excellent overview of such models. In its basic form, an I-O model is derived from a table of transactions, the elements of which represent flows of money between economic sectors. The table also includes columns, which represent demand in monetary terms from the different sectors of the economy, as well as rows, which represent employment, land and other types of inputs in the production process. Thus, a *transactions table* captures a snapshot of a national or regional economy at one point in time.

The advantage of a model based on a transactions table is that it captures the linkages between all the sectors of the economy. It also provides the means of estimating the level of employment and land required by each sector. In general, I-O models are demand driven, meaning that a change in demand for products from one or more sectors triggers a chain reaction, which stabilises at a new level of production, employment and land for each of the economic sectors involved. The heart of the model is a set of *technical coefficients*, one for each pair of sectors. Such a coefficient expresses the dollar value of input needed from one sector to produce one dollar's worth of output from the other sector. Once technical coefficients are estimated for one period in time, they remain fixed for the entire forecasting period. Thus, the I-O model accounts neither for economies of scale nor for technological change. Both these processes have the tendency to increase the value of the technical coefficients. This is the reason that the model is suitable for short-term forecasting (one to five years).

Several extensions to the model have been proposed over the years depending on the purpose of use. With regards to IUMs, the *multi-regional I-O model* has proven to be very useful (Leontief and Strout, 1963). Such a model deals with several regions simultaneously. It encompasses not only the linkages between sectors within each of the regions considered, but also the linkages of sectors across regions. The adaptation of such a model in an urban context is discussed in Macgill and Wilson (1979). The IUMs in Table 3.1 that make use of different types of multi-regional models are TRANUS and MEPLAN.

New Directions in Urban Modelling

In the early 1980s, planners realised that their efforts to reduce congestion and harmful environmental emissions through the provision of infrastructure did not achieve the desired results. Although many factors are responsible for these externalities, it is now recognised that reductions can only be achieved by increasing the efficiency and effectiveness of existing transportation facilities through *Travel Demand Management* (TDM) strategies. Such strategies fall into two categories: transportation and non-transportation alternatives (Plane, 1995). The former include high-occupancy vehicle (HOV) lanes, intelligent transportation systems (ITS) and congestion pricing, to name a few. In contrast, the latter alter people's travel behaviour through such strategies as alternative work schedules, telecommuting and jobs-housing balance. Many TDM strategies cannot be evaluated using existing IUMs because of their aggregate nature – that is, the basic unit of analysis is the zone. Strategies such as alternative work schedules can only be evaluated if the unit of analysis is an individual or a household. In other words, existing IUMs cannot be used to evaluate contemporary transportation policies. Thus, the future of urban modelling, as shown in Figure 3.2, lies with *behavioural IUMs*, which are conceptualised and implemented at a disaggregate level of analysis. The key component of such models is an *activity-based travel demand forecasting system.*

The remainder of this section summarises both empirical findings and recent modelling efforts in the field of activity analysis. The final part of this section discusses *object-oriented microsimulation* as a tool for implementing a behavioural IUM.

Activity-based Travel Analysis: Empirical Findings

In the late 1970s and early 1980s, travel behaviour research entered a new paradigm. *Activity-based travel analysis*, more commonly known as activity analysis, seeks to understand complex travel behaviour by recognising explicitly that travel is a demand, derived from the need to participate in out-of-home activities. Over the past two decades, activity analysis has contributed toward a better understanding of travel behaviour as evidenced by the proliferation of research in the field. More importantly, several activity-based models have been conceptualised and implemented for a one-day period. These models are likely to be the beginning of a new era of activity-based research in which the emphasis is on obtaining operational models that are useful in a planning context. Several reviews of activity-based travel analysis have been written at various times in the past 20 years, summarising the state-of-the-art in such research (Damm, 1983; Fox, 1995; Kitamura, 1988).

Activity-travel patterns Much activity-based research has focused on activity-travel patterns. Although the conceptualisation of an activity-travel pattern is consistent throughout most studies, researchers have used a variety of ways to classify such patterns, which vary in terms of complexity. For this chapter, an activity-travel pattern is defined as a sequence of out-of-home activities undertaken over a period of time such as a specific period of the day or the day itself. Two themes have emerged in this line of research: classification of activity-travel patterns and their association with explanatory factors and variability in activity-travel patterns over time.

Household lifecycle has played a key role in activity-based research because of the complex constraints that children impose on the activity-travel patterns of adults in the household. Kostyniuk and Kitamura (1982) examined the effect of this factor, along with household work-trip status and household role, on the evening activity-travel patterns of adult household members. Such patterns were classified according to whether they were made independently or jointly by household adults. The findings suggested that lifecycle stage of the household is related to many aspects of the evening activity-travel pattern, particularly its type – that is, whether or not adults participate in out-of-home activities in the evening, and if so, whether such activities are undertaken alone or together. Pas (1984) reports similar findings regarding the role of lifecycle in the activity-travel patterns of individuals over a one-day period.

Several researchers have investigated variability in activity-travel patterns over time. Hanson and Huff (1982, 1986, 1988) used the Uppsala (Sweden) Household Travel Survey to investigate the activity-travel behaviour of a sample of individuals over a 35-day period. The findings suggest that both repetition and variability characterise individuals' activity-travel patterns. However, the most important finding is that individuals exhibit more than one characteristic daily pattern. In related research, Pas and Koppelman (1986) investigate determinants of day-to-day variability in individuals' activity-travel behaviour. Their findings indicate that individuals who have fewer household and employment-related constraints exhibit more day-to-day variability in their activity-travel patterns. The work of Pas and Sundar (1995) support these earlier findings.

Activity-time allocation   Activity analysis adds a temporal dimension to travel behaviour research. Two areas have been investigated in this regard: the amount of time spent pursuing particular activity types over a period of time such as a day and the duration of a particular activity episode. The former area of research concerns *activity-time allocation*, whereas the latter concerns *activity episodes*.

Many researchers have attempted to explain individuals' daily allocation of time to different activities (e.g. Becker, 1965; Golob and McNally, 1997; Kitamura, 1984a; Levinson and Kumar, 1995). Others have investigated individuals' time allocation over longer periods (Kumar and Levinson, 1995). In these investigations, the activities are first classified. A common classification scheme is between *mandatory* and *discretionary activities*, which is most often a difference between work and non-work activities. For example, Kitamura (1984a) examined workers' out-of-home activity time allocation to non-work activities. The results indicated that auto-oriented individuals with a driver's license, more autos per driver and auto as the mode of work trip, tended to allocate more time to non-work activities. In contrast, work duration decreased the amount of time allocated to non-work activities.

Another classification scheme employed by Golob and McNally (1997) consisted of three activity types: work, maintenance and discretionary activities. This study was particularly insightful in that it considered household interactions between husbands and wives. The findings indicate that the amount of time allocated to out-of-home activities exhibited a hierarchy for both men and women – that is, work negatively affects the amount of time allocated to the other activity types and maintenance activity negatively affects the amount of time allocated to discretionary activity.

Activity episodes   An activity episode takes place within a measurable unit of time and is characterised by a uniform purpose and spatial setting. Over the course of a day, an individual participates in many activity episodes, which can occur in-home or out-of-home. For the most part, researchers have focused on out-of-home activity episodes, as they are the ones that generate travel. Many attributes of activity episodes have been investigated, including activity choice Kitamura and Kermanshah 1983), duration (Bhat, 1996a, 1996b; Ettema et al., 1995; Niemeier and Morita, 1996), destination choice (Kitamura, 1984b; Miller and O'Kelly, 1983) and sequencing (Kitamura, 1983; Kostyniuk and Kitamura, 1984; O'Kelly and Miller, 1984; Strathman et al., 1994). As well, some researchers have investigated several attributes jointly such as activity choice and duration (Bhat, 1998a; Damm, 1980, 1982; Damm and Lerman, 1981), activity choice and destination (Kitamura and Kermanshah, 1984), activity choice and home-stay duration (Mannering et al., 1994) and mode choice and number of episodes (Bhat, 1997). Finally, some researchers have reported results for several attributes – however, they are not investigated jointly (Hamed and Mannering 1993; Kitamura et al., 1997).

Kitamura and Kermanshah (1983) identify *time-of-day* and *history dependencies* of activity choice. For both home-based and non-home-based choice, time-of-day has a significant influence on activity engagement. History dependence has a more complex representation. Whether an individual has pursued an activity in the past influences his or her present activity choice in the home-based case. However, in the non-home-based case, only activities pursued during the current trip chain (a sequence of out-of-home activity episodes that begin and end at home) influence activity choice. These findings are supported by later work (Kitamura et al., 1997).

Hazard models are the primary analytical tools used by researchers to investigate the duration of activity episodes. With the exception of work by Ettema et al. (1995), the factors investigated are those that can be obtained from *trip* or *activity diaries*. Ettema et al. (1995) investigate factors that describe the process of activity scheduling. Their findings indicate that the time of day when the activity episode starts influences its duration. As well, factors such as the opening and closing times of activity sites and the priority that individuals assign to activity types influence activity episode duration. Similar to the findings for activity choice, the time spent in the same activity in the past influences its current duration.

A considerable body of literature has developed concerning *trip chaining*. This literature is related to activity analysis in that it concerns sequencing tendencies for out-of-home activities. Kitamura (1983) found that there exists a consistent hierarchical order in sequencing activities in that those activities that are less flexible tend to be pursued first. Strathman et al. (1994) show that the likelihood of forming complex commuting chains is higher for women, people who drive alone to work and workers from high-income households. Commuting during the peak periods shifts non-work activities away from the work commute. Household type is also shown to influence the propensity to form complex commuting chains, with single working adults with pre-school children having the highest propensity.

Activity scheduling Activity scheduling is the process concerned with the explicit choice, timing, duration, location and mode associated with activity episodes undertaken over a certain period of time such as a day or a week. This process has received increasing attention in recent years as researchers attempt to develop models of all-day activity-travel behaviour. Some researchers have investigated activity scheduling as a planning process whereby individuals select activities to undertake throughout the day (Ettema et al., 1994; Hayes-Roth and Hayes-Roth, 1979). The reason for such research is to identify heuristics that individuals use when scheduling their daily activities. Ettema et al. (1994) have shown that individuals plan activities in the order in which they are to be executed. The choice of activities, their location and sequence are found to be affected by the priorities that individuals assign them, as well as their duration, possible start and end times and travel times between locations.

Activity-based Travel Analysis: Modelling Efforts

Several activity-based models have been conceptualised and implemented for a one-day period. These models represent the first step towards the development of an activity-based travel demand forecasting system, which can be used in a behavioural IUM. This section summarises several aspects of these models as shown in

Table 3.2. These models differ in terms of their inputs, their objectives and the activity episode attributes that they capture.

Table 3.2 Characteristics of one-day activity-based models

Model	Key references	Agenda	Scheduling process	Attributes[11]
STARCHILD	Recker et al. (1986a, 1986b)	Yes		T, M
	Kawakami and Isobe (1990)			L
SMASH	Ettema et al. (1993, 1996)	Yes	Yes	T, L, C
SCHEDULER	Golledge et al. (1994)	Yes	Yes	T, L
	Ben-Akiva et al. (1996)			T, L, M, C
PCATS	Kitamura et al. (1996a)			T, L, D, M, C
AMOS	Kitamura et al. (1996b)	Yes		T, L, M
CATGW	Bhat (1998b)			T, D, L, M, C, R

[1] T = timing, D = duration, L = location, M = mode, C = activity type, R = travel time.

As shown in Table 3.2, several models require an *agenda* as input. An agenda is a list of activities, along with several attributes, that an individual may undertake over the course of a day. Agendas differ in their complexity. For example, STARCHILD requires information on the types of activities, their duration, their locations and any spatial, temporal and transportation constraints. In contrast, SMASH requires the following information for pre-defined activities: possible locations for performance, the number of times per day the activity can be performed, available time slots within which it can be performed at each location, duration, a measure of priority and the last time it was performed. What is apparent is that models that utilise an agenda require detailed information that is not collected in traditional travel or activity diaries. Additionally, from Table 3.2, there appears to be a relationship between the inclusion of an agenda and the number of activity episode attributes that are modelled explicitly. In some instances, although an attribute is listed, the agenda constrains the number of choices for it. For example, the locations for each activity modelled in SMASH are given in the agenda.

Table 3.2 also distinguishes between models that are designed to generate activity-travel patterns and those that are designed to replicate the cognitive process of activity scheduling. Once again, the models that are concerned with activity scheduling as a process require agendas as input and do not consider activity attributes explicitly.

In terms of developing an operational activity-based travel demand forecasting system for a behavioural IUM, the models in Table 3.2 that require an agenda present several problems. First, they require detailed data that are not collected in travel or activity diaries, which represent the state-of-the-practice in data collection. Second, modelling an agenda is a difficult task because many activity episode attributes are jointly determined such as activity choice and duration (Bhat, 1998a; Damm, 1980, 1982; Damm and Lerman, 1981). Finally, such models often use combinatorics, which is not practical when activity-travel behaviour must be generated for a large sample of individuals or even a population. The more practical approaches for an activity-based travel demand forecasting system are those that make use of existing data sources and model activity-travel patterns using existing statistical methodologies.

Object-oriented Microsimulation

The premier tool for operationalising behavioural IUMs is *microsimulation*. Miller (1996) defines microsimulation as an approach to modelling systems that are both dynamic – that is, they evolve over time – and complex in which the simulation model is formulated at the disaggregate level of individual decision-making units such as households, persons and firms. It is clear that activity-based travel demand forecasting systems require as input detailed socio-demographic information for households and individuals. In addition, information must be provided regarding the land-use and transportation characteristics of the urban area. The output from such a system is a detailed list of activities pursued by each individual on a given day, along with salient attributes, that can be used as input to traffic assignment algorithms. Microsimulation offers an attractive approach to evolving the decision-making units and characteristics of an urban area over time, therefore making medium- and long-term forecasting possible. Although travel behaviour researchers have used microsimulation in the past (e.g. Bonsall, 1982; Chung and Goulias, 1997; Goulias and Kitamura, 1992), its application to behavioural IUMs is virtually non-existent. In fact, only three conceptual frameworks have been proposed for such models: SMART (Simulation Model for Activities, Resources and Travel) (Stopher et al., 1996), SAMS (Sequenced Activity Mobility Simulator) (Kitamura et al. 1996b) and ILUTE (Integrated Land Use, Transportation, Environment modelling system) (Miller and Salvini, 1998). To date, none of these models have been operationalised for an urban area, although work is progressing towards this goal.

Microsimulation-based IUMs offer several advantages over traditional IUMs. They include data efficiency, behavioural interpretation, representation of complex interactions and mechanisms that affect behaviour, flexible aggregation of results and increased policy sensitivity. In the past, the primary disadvantages of microsimulation included data requirements and computational expense (Bonsall, 1982). Miller (1996) identifies a further concern. Although most microsimulation models work with a sample of the population, there are instances when it is necessary to use the entire population. This is the case for activity-based travel demand forecasting systems. The reason for this is that it is highly unlikely that the activity-travel behaviour of one household is representative of several households with similar characteristics, which is an implicit assumption when working with a sample of the population. Therefore, it is necessary to synthesise a population of decision-making units to be used as input to the microsimulation model. Two methods have been used for this task – that by Wilson and Pownall (1976) and that by Beckman et al. (1995). It is important to note that these disadvantages are no longer the limitations that they once were for two reasons: in both Europe and the United States, several activity-based data collection efforts have been undertaken and computing power is increasing at an exponential rate.

Perhaps the most important development that makes microsimulation increasingly attractive is the paradigm shift that occurred in the computing and information sciences in the early 1990s – that is, the shift from structured to object-oriented programming. *Object orientation* is more than simply an approach to computer programming – it encompasses a conceptual approach to modelling complex systems and is therefore well suited to the development of behavioural IUMs. Although it is beyond the scope of this chapter to discuss object orientation in depth, interested readers can consult any one of a number of excellent texts written on the subject (e.g. Booch, 1994; Booch et al., 1999; Rumbaugh et al., 1991).

Conclusions

The development of integrated urban models has now a history of over three decades. The initial excitement in the 1960s, fuelled by advances in information and computer technology, was followed by disappointment. The models developed were deemed to be too complicated and difficult to use with virtually no easy interface to planners and other non-expert users (Lee, 1973). In addition, the slowing down of metropolitan growth in the 1970s was associated with a general lack of interest in long-range planning. Thus, the 1970s and 1980s are characterised by slowing down in the effort to develop large-scale integrated urban models. Most advances during this period came from a relatively small number of researchers as summarised in Table 3.1.

In the 1990s, we see a renewed interest in IUMs. To a large extent, this interest is due to concerns raised by the worsening environmental conditions in urban areas, with the transportation sector identified as a major source of emissions harmful to human health. Reductions in emissions can be achieved not only through technological fixes, but also by reducing dependency on automobiles. Thus, transportation and land-use planners are required to co-ordinate their efforts in designing policies that will improve urban performance. Integrated urban models are thought to be effective tools in offering insights into such policies, thereby overcoming an important shortcoming of contemporary planning as identified in chapter 1.

Despite the relatively low profile of IUMs in the 1970s and 1980s, modellers gained significant experience in modelling social systems. The need to move into disaggregate models that encompass a more realistic representation of human behaviour became apparent. Random utility models, hazard models and advanced econometric estimation methods have now become an integral part of the urban modellers tool set. As discussed in the last section of this chapter, in the late 1970s and early 1980s considerable experimentation started with activity analysis. These ideas now find their way into models that attempt to simulate explicitly the behaviour of actors in the urban system. Accordingly, data collection efforts reflect these trends. Several metropolitan areas assemble and maintain disaggregated databases.

Unprecedented advances in computer technology support these efforts. Processing speed and storage capabilities are increasing at an exponential rate. New ideas in software engineering, such as object-oriented languages, are especially suited for the easy maintenance and incremental development of simulation programs.

Significant also is the widespread use of *Geographical Information Systems* (GIS), which provide an excellent platform for the development of IUMs. Such systems allow for the organisation and display of large volumes of spatial data, providing an excellent interface for experimentation and simulation scenario development by planners who are not trained in IUMs. Associated with GIS is also the development of *spatial statistics* over the last three decades. Only a handful of spatial statistical methods have found their way into commercial GIS programs. Such methods beyond statistical modelling also provide the means of matching spatial data from diverse sources, a useful capability for IUMs that often require large volumes of data.

References

Alonso, W. (1964), *Location and Land Use*, Harvard University Press, Cambridge.

Amano, K. and Abe, H. (1985) An Activity Location Model for the Metropolitan Area, *Infrastructure Planning Review*, 2, pp. 165-72.

Anas, A. (1983), *The Chicago Area Transportation Land-Use Analysis System*, Final Report, US Department of Transportation, Washington, D.C.

Anas, A. (1984), Discrete Choice Theory and the General Equilibrium of Employment, Housing, and Travel Network in a Lowry-type Model of the Urban Economy, *Environment and Planning A*, 16, pp. 1489-502.

Anas, A. (1992), *NYSIM (The New York Simulation Model): A Model of Cost-Benefit Analysis of Transportation Projects*, New York Regional Planning Association, New York.

Anas, A. (1995), METROSIM: A Unified Economic Model of Transportation and Land Use. Pamphlet distributed at the *Transportation Model Improvement Program's Land Use Modelling Conference*, Dallas (TX), February 19-21.

Anas, A. and Duann, L.S. (1986), Dynamic Forecasting of Travel Demand, Residence Location, and Land Development: Policy Simulations with the Chicago Area Transportation Land-use Analysis System, in B. Hutchinson and M. Batty (Eds.) *Advances in Urban System Modelling*, North Holland: Amsterdam, pp. 299-322.

Anderson, W.P., Kanaroglou, P.S. and Miller, E.J. (1994), *Integrated Land-Use and Transportation Model for Energy and Environmental Analysis: A Report on Design and Implementation*. Final Report, Ontario Ministry of Environment and Energy. Toronto, Ontario (Canada).

Anderson, W.P., Kanaroglou, P.S. and Miller, E.J. (1996a), Urban Form, Energy and the Environment: A Review of Issues, Evidence and Policy, *Urban Studies*, 33, pp. 7-35.

Anderson, W.P., Kanaroglou, P.S., Miller, E.J. and Buliung, R.N. (1996b), Simulating Automobile Emissions in an Integrated Urban Model, *Transportation Research Record*, 1520, pp. 71-80.

Barra, T., de la (1989), *Integrated Land Use and Transport Modelling: Decision Chains and Hierarchies*, Cambridge University Press, Cambridge.

Becker, G.S. (1965), A Theory of the Allocation of Time, *Economic Journal*, 75, pp. 493-517.

Beckman, R.J., Baggerly, K.A. and McKay, M.D. (1995), Creating Synthetic Baseline Populations, *Transportation Research A*, 30, pp. 415-429.

Ben-Akiva, M. and Lerman, S.R. (1985) *Discrete Choice Analysis: Theory and Application to Travel Demand*, MIT Press, Cambridge.

Ben-Akiva, M., Bowman, J.L. and Gopinath, D. (1996), Travel Demand Model System for the Information Era, *Transportation*, 23, pp. 241-266.

Bhat, C. (1996a), A Hazard-based Duration Model of Shopping Activity with Nonparametric Baseline Specification and Nonparametric Control for Unobserved Heterogeneity, *Transportation Research B*, 30, pp. 189-207.

Bhat, C. (1996b), A Generalised Multiple Durations Proportional Hazard Model with an Application to Activity Behaviour during the Evening Work-to-home Commute, *Transportation Research B*, 30, pp. 465-480.

Bhat, C. (1997), Work Travel Mode Choice and Number of Non-work Commute Stops, *Transportation Research B*, 31, pp. 41-54.

Bhat, C. (1998a), A Model of Post Home-arrival Activity Participation Behaviour, *Transportation Research B*, 32, pp. 387-400.

Bhat, C. (1998b), *A Comprehensive and Operational Analysis Framework for Generating the Daily Activity-travel Pattern of Workers*. Technical Paper, University of Texas at Austin, United States.

Bonsall, P.W. (1982), Microsimulation: Its Application to Car Sharing, *Transportation Research A*, 15, pp. 421-429.

Booch, G. (1994), *Object-Oriented Analysis and Design with Applications*, 2nd Edition, Addison-Wesley, Menlo Park (CA).

Booch, G., Rumbaugh, J. and Jacobson, I. (1999), *The Unified Modelling Language User Guide*, Addison-Wesley, Menlo Park (CA).

Bourne, L.S. (1982), Urban Spatial Structure: An Introductory Essay on Concepts and Criteria, in L.S. Bourne (Ed.) *Internal Structure of the City*, 2nd Edition, Oxford University Press, New York, pp. 28-45.

Boyce, D.E. and Lundqvist, L. (1987), Network Equilibrium Models of Urban Location and Travel Choices: Alternative Formulations for the Stockholm Region, *Papers of the Regional Science Association*, 61, pp. 93-104.

Boyce, D.E., Lupa, M.R., Tatineni, M. and He, Y. (1993), *Urban Activity Location and Travel Characteristics: Exploratory Scenario Analyses*. Seminar Paper, Environmental Challenges in Land Use Transport Co-ordination, December 6-10, SIG1, Blackheath.

Boyce, D.E., Tatineni, M. and Zhang, Y. (1992), *Scenario Analyses for the Chicago Region with a Sketch Planning Model of Origin-destination Mode and Route Choice. Final Report*, Department of Transportation, Illinois.

Brady, R.J. and McBride, J.M. (1992), *Technical Seminar on the ABAG Projection Process*. Special Conference of the Association of Bay Area Governments, Oakland (CA).

Brotchie, J.F., Dickey, J.W. and Sharpe, R. (1980), TOPAZ: Planning Techniques and Applications. In: *Lecture Notes in Economics and Mathematical Systems Series, Vol. 180*, Springer-Verlag, Berlin.

Bureau of Transport Economics (1998), *Urban Transport Models: A Review*. Working Paper 39, Canberra, Australia.

Caindec, E.K. and Prastacos, P. (1995), *A Description of POLIS. The Projective Optimisation Land Use Information System*, Working Paper 91-1, Association of Bay Area Governments, Oakland (CA).

CEC (1990), *Green Paper on the Urban Environment*, EUR 12902, Commission of the European Communities, Brussels.

Chung, J.-H. and Goulias, K.G. (1997), Travel Demand Forecasting Using Microsimulation: Initial Results from Case Study in Pennsylvania, *Transportation Research Record*, 1607, pp. 24-30.

Damm, D. (1980), Interdependencies in Activity Behaviour, *Transportation Research Record*, 750, pp. 33-40.

Damm, D. (1982), Parameters of Activity Behaviour for Use in Travel Analysis, *Transportation Research A*, 16, pp. 135-148.

Damm, D. (1983), Theory and Empirical Results: A Comparison of Recent Activity-based Research, in S. Carpenter and P. Jones (Eds.) *Recent Advances in Travel Demand Analysis*, Gower, Aldershot (UK), pp. 3-33.

Damm, D. and Lerman, S.R. (1981), A Theory of Activity Scheduling Behaviour, *Environment and Planning A*, 13, pp. 703-718.

Dickey, J.W. and Leiner, C. (1983), Use of TOPAZ for Transportation-land Use Planning in a Suburban County, *Transportation Research Record*, 931, pp. 20-6.

Echenique, M.H. (1968), *Urban Systems: Towards and Explorative Model*, Land Use and Built Form Studies, Cambridge.

Echenique, M.H., Flowerdew, A.D., Hunt, J.D., Mayo, T.R., Skidmore, I.J. and Simmonds, D.C. (1990), The MEPLAN Models of Bilbao, Leeds and Dortmund, *Transportation Reviews*, 10, pp. 309-22.

Ettema, D., Borgers, A. and Timmermans, H. (1993), Simulation Model of Activity Scheduling Behavior, *Transportation Research Record*, 1413, pp. 1-11.

Ettema, D., Borgers, A. and Timmermans, H. (1994), Using Interactive Computer Experiments for Identifying Activity Scheduling Heuristics. Paper presented at the 7*th* *International Conference on Travel Behaviour Research*, Santiago, Chile, June 13-16.

Ettema, D., Borgers, A. and Timmermans, H. (1995), Competing Risk Hazard Model of Activity Choice, Timing, Sequencing, and Duration, *Transportation Research Record*, 1493, pp. 101-109.

Ettema, D., Borgers, A. and Timmermans, H. (1996), SMASH (Simulation Model of Activity Scheduling Heuristics): Some Simulations, *Transportation Research Record*, 1551, pp. 88-94.

Floor, H. and de Jong, T. (1981), Testing a Disaggregated Residential Location Model with External Zones in the Amersfoort Region, *Environment and Planning A*, 13, pp. 1473-83.

Fotheringham, A.S. and O'Kelly, M.E. (1989), *Spatial Interaction Models: Formulation and Applications*, Kluwer Academic Publishers, London.

Fox, M. (1995), Transport Planning and the Human Activity Approach, *Journal of Transport Geography*, 3, pp. 105-116.

Garreau, J. (1991), *Edge City: Life on the New Frontier*, Doubleday, New York.

Golledge, R.G., Kwan, M.-P. and Gärling, T. (1994), Computational Process Modeling of Household Travel Decisions Using a Geographical Information System, *Papers in Regional Science*, 73, pp. 99-117.

Golob, T.F. and McNally, M.G. (1997) A Model of Activity Participation and Travel Interactions Between Household Heads, *Transportation Research B*, 31, pp. 177-194.

Goulias, K.G. and Kitamura, R. (1992), Travel Demand Forecasting with Dynamic Microsimulation, *Transportation Research Record*, 1357, pp. 8-17.

Hamed, M.M. and Mannering, F.L. (1993), Modelling Traveller's Postwork Activity Involvement: Toward a New Methodology, *Transportation Science*, 27, pp. 381-394.

Hanson, S. and Huff, J.O. (1982), Assessing Day-to-day Variability in Complex Travel Patterns, *Transportation Research Record*, 891, pp. 18-24.

Hanson, S. and Huff, J.O. (1986), Repetition and Variability in Urban Travel, *Geographical Analysis*, 18, pp. 97-114.

Hanson, S. and Huff, J.O. (1988), Systematic Variability in Repetitious Travel, *Transportation*, 15, pp. 111-135.

Hayes-Roth, B. and Hayes-Roth, F. (1979), A Cognitive Model of Planning, *Cognitive Science*, 3, pp. 275-310.

Hunt, J.D. (1993), A Description of the MEPLAN Framework for Land Use and Transport Interaction Modelling. Paper Presented at the 73*rd* *Annual Transportation Research Board Meetings*, Washington, D.C., January 9-13.

Hunt, J.D. (1994), *Calibrating the Naples Land Use and Transport Model*, Unpublished Paper, Department of Civil Engineering, University of Calgary, Calgary, Canada.

Hunt, J.D. and Simmonds, D.C. (1993), Theory and Application of an Integrated Land Use and Transport Modelling Framework, *Environment and Planning B*, 20, pp. 221-44.

Institute of Traffic Engineers (1987), *Trip Generation, 4th Edition.* Washington D.C.

Kanaroglou, P.S. and Anderson, W.P. (1997) Emissions from Mobile Sources in Urban Areas: An Integrated Transportation and Land-use Approach, in *Proceedings of the Fifth Conference of Environmental Science and Technology, Vol. A*, September 1-5, Molyvos, Lesvos, Greece, pp. 467-475.

Kawakami, S. and Isobe, T. (1990), Development of a One-day Travel-activity Scheduling Model for Workers, in P. Jones (Ed.) *Developments in Dynamic and Activity-Based Approaches to Travel Analysis*, Avebury, Aldershot (UK), pp. 184-205.

Kim, T.J. (1989), *Integrated Urban System Modelling: Theory and Practice*, Martinus Nijhoff, Norwell (MA).

Kitamura, R. (1983), Sequential, History-dependent Approach to Trip-chaining Behaviour, *Transportation Research Record*, 944, pp. 13-22.

Kitamura, R. (1984a), A Model of Daily Time Allocation to Discretionary Out-of-home Activities and Trips, *Transportation Research B*, 18, pp. 255-266.

Kitamura, R. (1984b), Incorporating Trip Chaining into Analysis of Destination Choice, *Transportation Research B*, 18, 67-81.

Kitamura, R. (1988), An Evaluation of Activity-based Travel Analysis, *Transportation*, 15, pp. 9-34.

Kitamura, R. and Kermanshah, M. (1983), Identifying Time and History Dependencies of Activity Choice, *Transportation Research Record*, 944, pp. 22-30.

Kitamura, R. and Kermanshah, M. (1984), Sequential Model of Interdependent Activity and Destination Choices, *Transportation Research Record*, 987, pp. 81-89.

Kitamura, R., Chen, C. and Pendyala, R.M. (1997), Generation of Synthetic Daily Activity-travel Patterns, *Transportation Research Record*, 1607, pp. 154-162.

Kitamura, R., Fuji, S. and Otuka, Y. (1996a), An Analysis of Induced Travel Demand Using a Production Model System of Daily Activity and Travel, which Incorporates Space-time Constraints. Paper presented at the *5th World Congress of the Regional Science Association International*, Tokyo, Japan, May 2-6.

Kitamura, R., Pas, E.I., Lula, C.V., Lawton, T.K. and Benson, P.E. (1996b), The Sequenced Activity Mobility Simulator (SAMS): An Integrated Approach to Modelling Transportation, Land Use and Air Quality, *Transportation*, 23, pp. 267-291.

Kostyniuk, L.P. and Kitamura, R. (1982), Life Cycle and Household Time-space Paths: Empirical Investigation, *Transportation Research Record*, 879, pp. 28-37.

Kostyniuk, L.P. and Kitamura, R. (1984), Trip Chains and Activity Sequences: Test of Temporal Stability, *Transportation Research Record*, 987, pp. 29-39.

Kumar, A. and Levinson, D. (1995), Temporal Variations on Allocation of Time, *Transportation Research Record*, 1493, pp. 118-127.

Lee, D.A. (1973), Requiem for Large-scale Models, *Journal of the American Institute of Planners*, 39, pp. 163-178.

Leontief, W.W. (1941), *The Structure of the American Economy 1919-1939*, Oxford University Press, New York.

Leontief, W.W. and Strout, A. (1963), Multi-regional Input-output Analysis, in T. Barna (Ed.) *Structural Interdependence of Economic Development*, MacMillan, London, pp. 119-149.

Levinson, D. and Kumar, A. (1995) Activity, Travel, and the Allocation of Time, *Journal of the American Planning Association*, 61, pp. 458-470.

Lowry, I.S. (1964), *A Model of Metropolis*, RM-4053-RC, Rand Corporation, Santa Monica (CA).

Macgill, S.M. and Wilson, A.G. (1979), Equivalences and Similarities Between Some Alternative Urban and Regional Models, *Systemi Urbani*, 1, pp. 9-40.

Mackett, R.L. (1983), *The Leeds Integrated Land Use-Transport (LILT) Model*, Supplementary Report 805, Transport and Road Research Laboratory, Crowthorne (UK).

Mackett, R.L. (1990a), The Systematic Application of LILT Model to Dortmund, Leeds and Tokyo, *Transportation Reviews*, 10, pp. 323-338.

Mackett, R.L. (1990b), Comparative Analysis of Modelling Land Use-transport Interaction at the Micro and Macro levels, *Environment and Planning A*, 22, pp. 459-475.

Mackett, R.L. (1990c), *MASTER Model (Micro-Analytical Simulation of Transport, Employment and Residence)*, Report SR 237, Transport and Road Research Laboratory, Crowthorne (UK).

Mackett, R.L. (1991a), A Model-based Analysis of Land Use and Transport Policies for Tokyo, *Transportation Reviews*, 11, pp. 1-18.

Mackett, R.L. (1991b), LILT and MEPLAN: A Comparative Analysis of Land Use and Transport Policies for Leeds, *Transportation Reviews*, 11, pp. 131-154.

Mannering, F., Murakami, E. and Kim, S.-G. (1994), Temporal Stability of Traveller's Activity Choice and Home-stay Duration: Some Empirical Evidence, *Transportation*, 21, pp. 371-392.

Martinez, F.J. (1997a), Towards a Microeconomic Framework for Travel Behaviour and Land-use Interactions. Resource Paper, *8th Meeting of the International Association of Travel Behaviour Research*, Austin (TX), United States.

Martinez, F.J. (1997b), *MUSSA: A Land Use Model for Santiago City*, Department of Civil Engineering, University of Chile, Santiago.

Martinez, F.J. and Donoso, P.P. (1995), MUSSA Model: The Theoretical Framework, in *Proceedings of the 7th World Conference on Transportation Research*, Sydney, Australia.

McFadden, D. (1973), Conditional Logit Analysis of Qualitative Choice Behaviour. In: P. Zaremka (Ed.) *Frontiers in Econometrics*, Academic Press, New York, pp. 105-142.

Miller, E.J. (1996), Microsimulation and Activity-based Forecasting, in Texas Transportation Institute (Ed.) *Activity-Based Travel Forecasting Conference, June 2-5, 1996: Summary, Recommendations, and Compendium of Papers*, Travel Model Improvement Program, U.S. Department of Transportation and U.S. Environmental Protection Agency, , Washington, D.C., pp. 151-172.

Miller, E.J. and O'Kelly, M.E. (1983), Estimating Shopping Destination Choice Models from Travel Diary Data, *Professional Geographer*, 35, pp. 440-449.

Miller, E.J. and Salvini, P.A. (1998), The Design and Evolution of an ILUTE Dynamic Microsimulation Framework. Paper presented at the *77th Annual Meeting of the Transportation Research Board*, January.

Miller, R.E. and Blair, P.D. (1985), *Input-Output Analysis: Foundations and Extensions,* Prentice-Hall, New Jersey.

Nakamura, H., Hiyashi, Y. and Miyamoto, K. (1983), Land Use-transportation Analysis System for Metropolitan Area, *Transportation Research Record,* 931, pp. 11-20.

Niemeier, D.A. and Morita, J.G. (1996), Duration of Trip-making Activities by Men and Women, *Transportation,* 23, pp. 353-371.

O'Kelly, M.E. and Miller, E.J. (1984), Characteristics of Multistop Multipurpose Travel: An Empirical Study of Trip Length, *Transportation Research Record,* 976, pp. 33-39.

Papageorgiou, Y.Y. and Pines, D. (1999), *An Essay on Urban Economic Theory,* Kluwer Academic., Norwell (MA).

Pas, E.I. (1984), The Effect of Selected Sociodemographic Characteristics on Daily Travel-activity Behaviour, *Environment and Planning A,* 16, pp. 571-581.

Pas, E.I. and Koppelman, F.S. (1986), An Examination of the Determinants of Day-to-day Variability in Individuals' Urban Travel Behaviour, *Transportation,* 13, pp. 183-200.

Pas, E.I. and Sundar, S. (1995), Interpersonal Variability in Daily Urban Travel Behaviour: Some Additional Evidence, *Transportation,* 22, pp. 135-150.

Plane, D.A. (1995), Urban Transportation: Policy Alternatives, in S. Hanson (Ed.) *The Geography of Urban Transportation, 2nd Edition,* New York, Guilford, pp. 435-469.

Prastacos, P. (1986a), An Integrated Land Use-Transportation Model for the San Francisco Region: 1. Design and Mathematical Structure, *Environment and Planning A,* 18, pp. 307-22.

Prastacos, P. (1986b), An Integrated Land Use-Transportation Model for the San Francisco Region: 2. Empirical Estimation and Results, *Environment and Planning A,* 18, pp. 511-28.

Putman, S.H. (1983), *Integrated Urban Models: Policy Analysis of Transportation and Land Use,* Pion, London.

Putman, S.H. (1991), *Integrated Urban Models 2: New Research and Applications of Optimization and Dynamics,* Pion, London.

Recker, W.W., McNally, M.G. and Root, G.S. (1986a), A Model of Complex Travel Behaviour: Part I – Theoretical Development, *Transportation Research A,* 20, pp. 307-318.

Recker, W.W., McNally, M.G. and Root, G.S. (1986b), A Model of Complex Travel Behaviour: Part II – An Operational Model, *Transportation Research A,* 20, pp. 319-330.

Rumbaugh, J., Blaha, M., Premerlani, W., Eddy, F. and Lorensen, W. (1991), *Object-Oriented Modelling and Design,* Prentice-Hall, New Jersey.

Scott, D.M., Kanaroglou, P.S. and Anderson, W.P. (1997), Impacts of Commuting Efficiency on Congestion and Emissions: Case of the Hamilton CMA, Canada, *Transportation Research D:* 2, pp. 245-257.

Sharpe, R. (1978), The Effects of Urban Form on Transport Energy Patterns, *Urban Ecology,* 3, pp. 125-35.

Sharpe, R. (1980), Improving Energy Efficiency in Community Land Use-Transportation Systems, *Environment and Planning A,* 12, pp. 203-16.

Sharpe, R. (1982), Local Government Conservation Using Computer Modelling, *Journal of Royal Australian Planning Institute*, 20, pp. 118-20.

Sheffi, Y. (1985), *Urban Transportation Networks: Equilibrium Analysis with Mathematical Programming Methods*, Prentice-Hall, New Jersey.

Southworth, F. (1995), *A Technical Review of Urban Land Use–Transportation Models as Tools for Evaluating Vehicle Travel Reduction Strategies*, Report ORNL-6881, Oak Ridge National Laboratory, Oak Ridge (TN).

Stopher, P.R., Hartgen, D.T. and Li, Y. (1996), SMART: Simulation Model of Activities, Resources and Travel, *Transportation*, 23, pp. 293-312.

Strathman, J.G., Dueker, K.J. and Davis, J.S. (1994), Effects of Household Structure and Selected Travel Characteristics on Trip Chaining, *Transportation*, 21, pp. 23-45.

Waddell, P. (1998a), The Oregon Prototype Metropolitan Land Use Model, in *Proceedings of the ASCE Conference on Transportation, Land Use and Air Quality: Making the Connection*, Portland, Oregon.

Waddell, P. (1998b), An Urban Simulation Model for Integrated Policy Analysis and Planning: Residential Location and Housing Market Components of URBANSIM. Paper presented at the *8th World Conference on Transport Research*, Antwerp, Belgium.

Watterson, W.T. (1993), Linked Simulation of Land Use and Transportation Systems: Developments and Experience in the Puget Sound Region, *Transportation Research A*, 27, pp. 93-206.

Wegener, M. (1982a), A Multilevel Economic-demographic Model for the Dortmund Region, *Sistemi Urbani*, 4, pp. 371-401.

Wegener, M. (1982b), Modelling Urban Decline: A Multilevel Economic-demographic Model for the Dortmund Region, *International Regional Science*, 7, pp. 217-241.

Wegener, M. (1986), Transport Network Equilibrium and Regional Deconcentration, *Environment and Planning A*, 18, pp. 437-56.

Wegener, M. (1994), Operational Urban Models: State of the Art, *Journal of the American Planning Association*, 60, pp. 17-29.

Wegener, M. (1995), Current and Future Land Use Models, in *Land Use Modelling Conference Proceedings*, Arlington (TX).

Wilson, A.G. (1970), *Entropy in Urban and Regional Modelling*, Pion, London.

Wilson, A.G. (1974), *Urban and Regional Models in Geography and Planning*. Chichester: John Wiley and Sons.

Wilson, A.G. and Pownall, C.E. (1976), A New Representation of the Urban System for Modelling and for the Study of Micro-level Interdependence, *Area*, 8, pp. 246-254.

Wilson, A.G., Coelho, J.D., Macgill, S.M. and Williams, H.C.W.L. (1981), *Optimisation in Locational and Transport Analysis*, John Wiley and Sons, Chichester.

4 Understanding the Political and Administrative Framework of Urban Performance

ERKKI MENNOLA

Making European Cities Comparable

The goals of the COST-CIVITAS programme are ambitious. The results should help us understand and interpret new urban challenges for Europe and provide a point of view for urban policies. The knowledge should be transferred from one city, country and culture to another, whether this information concerns the entire European urban system or only one city or type of city. The programme should help European cities advance economically and socially.

The initiators of the programme seem to have been well aware of the problems caused by the academic specialisation and the conventional border lines between disciplines. Academics and policy-makers have trouble agreeing on urban questions. This lack of agreement concerns the social and economic significance of cities and urban networks in terms of development, production of wealth and power, as well as their social and geographical nature. The divergence is due to the fact that there is no global scientific approach to understanding European urban systems. Our particular mission has been to formulate the general European aims of the COST-CIVITAS programme with respect to the management of urban infrastructure and public services as far as their economic

73

efficiency and social effectiveness are concerned. Strong arguments, particularly the limited research resources and the usability of the results, led us to take a pragmatic and concentrated approach. As a result, we now have a rich collection of case studies, presenting interesting planning and regeneration projects for major infrastructure and services in several European cities. We also have specific theoretical tools to analyse these cases from the performance of urban point of view.

However, in view of the general aims of the programme, the question of the *European understanding* of the urban problems should be raised simultaneously. Obviously there is always a danger that single cases will remain obscure and specific theories will be misunderstood and misused in the absence of a broader contextual framework and knowledge base. People coming from different countries and cultures, with various academic and professional competencies, make selective use of the message. On the other hand, even the local experts might see their projects in a new light and understand their results when they are placed in a wider European context.

No specific discipline can bring us the needed comprehensibility and comparability. No academic approach can cover more than a narrow slice of life, out of the innumerable physical and human factors producing the interactive complex system called a city. In addition to the specialised scientific theories, something more is needed to put things into the right context (Wachter, 1995).

What we need at this point are theoretical links to connect single-case studies and specific theoretical considerations with the general political, administrative and legal environment and evolution of cities in Europe. I shall try to give an answer (or more precisely, some helpful guidelines for giving an answer) to two important questions. These always seem to come up when we are dealing with a case study in one or several cities and interpreting its results. The questions are:

- does this approach cover every relevant component of a city in this *context*?
- are the results *comparable* with those from other European cities in their specific political and social environment?

We should always know the relevant properties of the region concerned in a wider context in order to understand the message of the case, to join in a scientific or political discourse based on it, or to utilise the results of the research project. The next section concerns the features that count for our context (urban performance and production of infrastructure services). There I explain why precisely these features are so important to us. I also make an effort to reduce the overwhelming amount of details to manageable models. The typologies I suggest include the relevant dimensions of a European city as a political, administrative and legal entity.[1]

The Components of the Political and Administrative City

The Problem of Complexity

First of all, we must be aware that we are working with perhaps the *most complex institution* ever created. A city has far more dimensions than, for instance, a national state, ranging from multinational economic operations to the smallest details of the natural environment. And its complexity is only increasing, as we shall see.

A city has both *horizontal* and *vertical* dimensions. A city's time dimension, its *history*, is of particular importance when evaluating and comparing the political environment.[2] The modern European city, both as an urban structure and as a political and administrative system, is always a result of the development of hundreds of years at least. Because of its unique historical background, every city is an individual case. Yet we can also identify certain periods of the life cycle of all cities when they have been more or less influenced by the same universal forces. It is highly probable that our own Internet Age will be one such period.[3]

First and foremost, cities are *human* institutions. They include all those complexities that attend human life. The human nature of organisations means that things like traditions, feelings, incentives, communication, creativity and responsibility play a more or less prominent role in all activities. Cities are in fact combinations and networks of innumerable actors. The actors are individual citizens and various public and private bodies, all of them with goals and views of their own. The complexity of a city is exponential compared with its components, which in themselves may already be very complex.[4]

One of the most important properties of a city as a human organisation is continuous *change*. Millions of actors – both privately and in various public activities, individually or as organisations – are working *inside* the cities to make some kind of change in their individual sphere of life. In addition, global influences coming from the *outside*, particularly through the globalisation of business life, are increasingly causing local changes everywhere in the world. It is sometimes said that the most important local decision-maker in a contemporary European city may be a New York stockbroker in his late twenties.

Understanding and managing cities, in their political and administrative role as well as concerning any other aspect presupposes a *holistic* view. One must be able to incorporate and simultaneously process things that normally do not belong together by using methods that the established sciences do not serve as given.[5] The task lying ahead of us now has sometimes been compared to the leap from mechanistic Newtonian physics into quantum physics. The allegories concerning both the problems and the potential solutions are interesting indeed (Zohar and Marshall, 1993).

The efforts to deal with the political and administrative environment and the framework of *local authorities* holistically by scientific means is not a modern phenomenon. We can find expressions of it among the philosophers of Antiquity. There was a very interesting period in the history of science before World War I when attempts were made – particularly in Germany (*Kommunalwissenschaft*), but also in the USA (Municipal Science) – to develop a specialised science to study local authorities and local self-government as a combination of political, administrative, economic, social, cultural and legal functions. The rapidly advancing specialisation and diversification of sciences as well as the political events in the countries concerned undermined those efforts, however.[6]

A sound approach would obviously be an optimal combination of large-scale theoretical considerations and empirical observations of details. We should keep in mind the statement by C. Wright Mills

(1980) that bad researchers are "both those who observe without thinking and those who think without observing".

When observing 'facts' concerning different properties and functions of cities, we must also remember that just a limited share of them are 'brute' facts that do not need human institutions to be proved. Quite many of them, like for instance 'government' or 'money', are what John R. Searle (1995) calls *institutional facts*. Actually, they are just human agreements. When working with local institutions, we must remember that many of the human institutions we meet there are national or even regional constructions and agreements and may vary from country to country.

The ontological vulnerability and uncertainty of our holistic considerations are obvious, of course. We are dealing with problems that belong to the most advanced level of human life.[7] Anyway, we should try to bring our detailed observations and analyses up to this complex level, too, following the advice of St. Thomas Aquinas, who said (Summa theologica, I, 1, 5 ad.1) that: "The slenderest knowledge that may be obtained of the highest things is more desirable than the most certain knowledge obtained of lesser things".

It is a mistake to separate culture and nature in any scientific approach. In our complex field, the problem of reduction could lead to exceptional difficulties. Marjorie Grene (1987) has presented two important principles that I use as guidelines. Her key concepts are *historical reality* and *contextual objectivity*; this involves putting things into the right perspective concerning time and importance. As she says, the question is always *what needs explaining in the context of the current problem*, and what other concepts or principles or methods can appropriately be imported here and now from other disciplines.

Political Geography

The socio-geographical properties of a city play a considerable role even where its political, administrative and legal existence is concerned. The physical, economic and cultural dimensions are important. All of them have an influence on the *political geography* of the city (Urwin, 1990).

All cities, at least those considered as metropolitan regions, have at least three geographical levels of political and administrative functions: the tiers of *neighbourhood community*, *agglomeration* and *economic region.* They reflect the social, physical and economic dimensions of the city. The organisational state of affairs of these three functional levels will strongly influence many local activities (Mennola, 1991 and 1998).

The political geography of the city is implemented by the *territories* of local authorities and other units. The official political and administrative map and the principles of geographical division of the existing local authorities may deviate from the functional tiers mentioned above. The territorial structure of local government around Europe varies considerably. It is not unusual to find a city where none of those three functional tiers is geographically institutionalised. That is particularly the case in Northern Europe, where large amalgamations of local authorities have been typical since World War II (Jensen, 1994; Delcamp, 1990).

Even political geography is a result of a long *history.* There has been physical growth, rapid industrialisation and administrative reforms but also, in some cases, backlashes and decline, all bringing specific features to the existing political and administrative map. This process has by no means come to an end. Particularly at the agglomeration level, many European cities, as well as the economic regions around them, are experiencing a period of strong change at the moment (Council of Europe/CDLR, 1993).

The roots of political geography in cities are mainly regional or national. But even models and ideas of international origin have played a decisive role in the history of local government. For instance, the prevailing structure of numerous small municipalities in Latin Europe originated in the Napoleonic administration. And the British local government boundary reforms after World War II gave impetus to similar actions in many North European countries (Delcamp, 1990; Rowat, 1980).

The political geography of a city is one of the potential variables having great influence on urban performance. Drawing conclusions and generalising from observations on one city to the European level presupposes comparative information and knowledge about the political map of the cities concerned. It is important to have an overview of the tiers of government, administrative divisions, constituencies, historical roots, traditions and culture of a city.

Autonomy and Democracy

Local self-government is one of the basic institutions of European democracy. That is why it is considered a prerequisite of membership in many European organisations. When comparing European cities, we can assume that we are going to meet some application of the principle of local self-government everywhere. In other parts of the world this is not always the case (Council of Europe/CLRAE, 1980 and 1981; Leemans, 1970).

The local application of self-government may vary, however. There are national and, in the federal or semi-federal countries, even regional differences. We can also define more or less uniform European *cultural zones of local self-government*, crossing national borders in different parts of the continent. Outside Europe, for instance the USA and Canada have their own cultures of local self-government.

The emergence of specific cultural zones with numerous similar features in their systems is again due to a long history. It is a combination of legal, political and cultural traditions, geography, economic structure and cultural hegemonies. Researchers have defined between two and six different cultures of local self-government in Europe, depending on the criteria applied.[8]

The basic line of demarcation runs between the *North* and the *South*, approximately from Belgium to Northern Italy. It is amazingly analogous with the culture boundary in many other things, sometimes called the 'soft cheese-hard cheese' or 'wine-beer' line. The historical roots of this division can be found in many sources, from the ancient Roman empire and the Roman law to Catholic and Protestant religions and the extents of the Napoleonic regime. There are relevant differences within the northern zone between the *British*, the *Central European* and the *Scandinavian* self-government cultures, however. All of them can be considered individual zones as well. The emerging democracies of *Eastern and Central Europe* constitute a category of their own, although they do not represent any uniform culture of local self-government (ILGPS, 1994).

When analysing the properties of the various European applications of the idea of local self-government, there is now also available a largely respected common standard, the *European Charter of Local self-government*.[9] It was approved in October 1985 by the Council of Europe, after a preparatory process of decades, and was then opened for signatures to become an international treaty. At the moment, 36 European countries have already ratified or signed the treaty.

The European Charter of Local self-government is an important yardstick in two respects concerning comparative research of local authorities and their activities. Firstly, it gives the *minimum criteria* for an acceptable solution to local self-government in any European country. Secondly, it expresses the *relevant properties* of any system of local self-government. We can start from the presumption that at least all those points of view that have been included in the Charter are important to the status and competencies of any local authority in a European democracy.

Those criteria help us define the political status of the various tiers of metropolitan or city government, too. Even if there is a specific organisation for administering the agglomeration level of the city, it may not be a single political unit of local self-government but just an association of numerous independent municipalities. Correspondingly the sub-units of a city (neighbourhood councils, etc.), often are just boards or divisions of the local authorities. Criteria set by the Charter, like having elected decision-makers, independent budget and freedom to decide about their internal administrative structures, reveal the relevant differences concerning their position.

Amongst the various elements of local self-government, the *financial autonomy* of local authorities is of particular importance concerning major infrastructure investments and services. The local tax systems, grants from the national government and other sources of income of local authorities vary largely in Europe. That is the case concerning freedom to use their resources, too. However, the Charter also guarantees some basic rights of the local authorities to exercise their own political discretion. The competencies of the local and regional authorities in obtaining money from the international financial market may be strongly limited in some countries. In contrast, some other countries have not set any legal limits on local government borrowing (Council of Europe/CLRAE, 1990 and 1991; Council of Europe/CDLR, 1992).

The system of *political management* in a city may influence decisions on major infrastructure projects. Politically strong, sometimes even directly elected mayors are typical of Southern European cities. It is probable that decision-making on large and expensive investments is more single-minded there, at least as far as the local authority itself is concerned, than in Northern European local government. Northern European cities traditionally favour proportionality and more collective forms of decision-making. If parliamentary principles and majority rule are applied even to local government – as is the case in England and in the Netherlands as well as in Oslo, Norway – political decision-making has yet another type of environment.

Members of a political community like a city represent a whole spectrum of opinions and interests. The political decision-making in a democracy, with all its pressure groups, reflects this diversity of views. Decisions on major infrastructure and services in a city always contain many types of (even controversial) political issues. What is actually the best solution may even remain a permanently open question. Researchers may present powerful arguments in favour of or against some alternatives, but they cannot solve political controversies based on values. Those belong to the political field.

We can see that European comparisons of major infrastructure services presuppose an understanding of political structures. We must know how the projects are initiated and how political and financial decisions are made.

Administration and Activities

The diversity of administrative and other functional structures in European cities is perhaps even more striking. True, basic infrastructure and services are more or less similar everywhere. Even the standard of those services may be largely comparable, at least amongst the EU member states. However, the way those services and functions have been organised, may vary considerably from country to country (Delcamp, 1994).

An important difference in this context is the division of responsibilities between the *public* and the *private* sector. Certain European political cultures, for instance the British, have traditionally favoured solutions operated by the private sector. On the other hand, particularly the Scandinavian welfare states have constructed a large public sector where local government is – or at least has been – the main producer of various services in cities (Hesse and Sharpe, 1991).

Within the publicly operated service sector we can see that in Southern Europe, many local services have traditionally been provided by the *state* or its regional bodies. In the North, the self-governing *local authorities* usually have been in charge of them. Particularly the administrative differences concerning the organisation of such voluminous local services like health care or education make the Northern European municipalities look quite different from their Southern counterparts (Batler and Stoker, 1991).

Major infrastructure services show a wide variety of approaches that are not in line with our conventional expectations. For instance the local energy supply in Europe includes all possible forms of organisation, from purely local-government operations to multinational corporations. Local bus traffic may be operated by the local authority concerned, by a regional or national publicly owned company, or by a private enterprise. The same multiplicity is found today in any branch of local service provision.

In spite of the innumerable differences and controversies, we can see parallel developments. The trend goes *from public to private* and *from national to multinational* operators in organising local services. Such tendencies are particularly obvious concerning services of a more or less technical nature. There is a historical rather than a rational background to the phenomenon that certain local services like electricity delivery, water supply, garbage disposal or public transport have traditionally been regarded as public responsibilities. Basically, of course, they are just that kind of prerequisites of urban living that the town bourgeoisie has historically wanted to manage independently by means of their local self-government. Today these services are increasingly regarded as normal industrial activities, and they are going to be organised accordingly (Mennola, Morange and Kinnala, 1995).

Today, it is European integration and the *European single market* that is propelling the organisational homogenisation of industries like these. The EU is not regulating the way the member states organise their local government. Yet the EU legislation, promoting free movement of people and capital, opening closed markets for competition, unloading monopolies, homogenising norms, subjecting local authorities to compulsory tendering etc., is strongly promoting the liberalisation of local service provision. For the coming years, this will mean more private activities and more international enterprises in any branch. An important but often unnoticed form of development inside public-sector institutions is the so-called back-door privatisation, advancing along with the rapidly increasing subcontractor networks.[10]

On the other hand, the private sector is becoming more and more responsible for the operational service provision in cities. Nonetheless these services are and will always be a local affair and a *concern of local politics*. It is of utmost importance to the local population and economy to know how the services are organised and to be able to influence their implementation. New instruments for safeguarding local will probably be developed, from project management and co-finance to concessions and control functions.

It is thus necessary for a researcher to obtain extensive knowledge of the variety of ways in which service provision is organised in European cities. The researcher should also understand current tendencies, even when analysing a single local project or one detail of local services. It is particularly important to monitor the impact of European integration on local service provision.

Legal Framework

The legal system is the *method to implement* the political geography, the local autonomy and democracy, and the organisation of public administration and services. The definition of the 'institutional facts' mentioned above is published in the constitution, legislation and other legal instruments.

The legal system always plays an important, sometimes even a decisive role in the complex economic and social system called a city. Competencies, responsibilities and territories of the local authorities and their co-operators and counterparts are based on law. The same is true of the political and administrative processes and actors concerned. Law provides the framework for planners, sets the standards, and tells what is allowed and what is not. A legal system is thus an interactive and dynamic part of the whole, influencing the other elements and receiving influence from them.

The dynamic role of the legal system is often neglected by researchers of society and the economy, however. The reason is that scientists studying law and those in other fields unfortunately do not communicate with each other very much. The traditional scientific interests of the law scholars have been in interpreting and systematising existing legal systems, not in understanding the change (Mennola, 1998 b).

Integrated studies on cities, including a contribution from the legal expertise as well, would be valuable. They could deepen our understanding of the role of law and the legal system as a part of the city dynamics. A systematic approach is needed, if only because of the large number of legal constructions and instruments. In many European countries, local government, planning, construction, business regulations, technical norms, and many other preconditions of local infrastructure and services are competencies of *regional legislation*. This means that there are actually hundreds of more or less individual legal subsystems in our European regions and cities.

It is obvious that the legal framework of local governance will become even more complex in the future. In addition to the increasing local and regional competence to set norms and regulations, we also have *EU legislation*. Directly and indirectly, it brings new European elements to the legal framework where we act locally.

We shall thus not ignore the legal system as a relevant framework of urban performance and infrastructure. To understand the whole, we need some legal knowledge, too, both of the principles of European local government in general and of the local structures concerned.

Managing Complexity, Context and Change

Approval of Limited Goals

The limits of applying scientific approaches are particularly obvious when trying to understand and manage a complex human institution like a city. Here, we briefly summarise the kinds of problems the researcher is likely to meet on the way:

- the elements of local and regional structures and institutions have an impact on each other. Changes in any of them may cause changes in the whole system;
- the contemporary structures and status of local and regional institutions are the result of a unique long-term historical development. These institutions are undergoing continuous change;
- the number of details potentially relevant to the system is huge and beyond the capacity of human knowledge to comprehend. No single researcher or project can hope to cover more just a negligible share of those things;
- the problems deal with too many fields of scientific expertise to be assigned to any working group or other single organ;
- the knowledge needed is partly non-conceptual in nature, presupposing personal experiences from the cities and cases concerned. The difficulty of obtaining such knowledge is obvious.

I would like to suggest adopting a very humble attitude in the face of the overwhelming challenge to understand and manage the problems of the urban performance and infrastructure by scientific means. A researcher can make a positive contribution, but can also get completely lost. Contemporary phenomena are often misleading; the real forces in play and their results may be invisible today. That is why seeking guidance from more *timeless problems and historical examples* is often useful.

Because of the long-standing trend toward further specialisation in the sciences, we must go back to the age when scientists were expected to be broadly based scholars. *Max Weber* provided very useful, let us say even modern, tools to understand and to manage cities, even in their utmost complexity. His constructs are particularly valuable for our purposes because they allow us to combine quantitative and qualitative, as well as empirical and more philosophical approaches.

The Ideal Types of Max Weber

The key concept formulated by Max Weber[11] is the *ideal type*. He constructed several ideal types of towns: the merchandise town, government town, manufacturing town, leisure and consumption town, and military town. Then he used these ideal types to describe and interpret his views on such towns as compositions of many different factors. Amongst these factors, he considered the political, economic, legal and religious properties to be the most important. He strongly opposed attempts by some of his contemporaries to reduce towns and cities to simple economic constructs. To Weber, a town was definitely not just any built-up area or concentration of population. Rather, it was a community that met numerous criteria, both physical and human. [12]

Constructing an ideal type, understood as Weber used the concept, means *(over)emphazing* one or several points of view to reveal the typical character of the city. This entails *collecting* a number of diffuse and discrete single phenomena, appearing more or less in this context. Finally, it involves drafting an *overall picture* based on these observations.

Weber never intended an ideal type to be a description of reality but as a kind of *defined utopia*. It was meant to play the same role as the hypothesis in the natural sciences. Thus it can be used when certain non-relevant features are to be left outside the observation. An ideal type is actually a *question* posed to the future empirical research. That research is going to give its reply in time, either to confirm or to reject our proposed interpretation. *'Die Stadt'* itself as a book is a powerful argument in support of the method developed by Weber.

Weber used his ideal types of towns to explain the process that led to the birth of modern democracy and the market economy. He showed why the properties that made Europe the leading continent were emerged here and not, for instance, in some of the equally or even more developed empires in the Near or Far East. The economic performance of the European towns and cities was his main argument in explaining the causalities. In brief, their performance was due to their political and legal structures and the role of the emerging European urban bourgeoisie as a new independent political and economic actor.

Weber's method of ideal types is useful even today, and perhaps more than ever. It helps explain the differences in performance among our cities. And it is useful when analysing the political and administrative environment of various local activities, functions and projects.

Below, I attempt to adapt Weber's ideal types to analyse how our cities are governed and administered. My purpose is to find a holistic starting point for our efforts to analyse urban performance in major infrastructure services.

Ideal Types of City Governance

The central idea of the Weberian ideal types is, analogous to my considerations in section 1, that cities are composed of several *essential interactive elements*. As an ideal type, the city must always be understood as a *combination* of these different qualities, not as a single-issue institution. Understanding the special character of a given city presupposes a comprehensive simultaneous observation of all the relevant elements. Changes in any of them may have an effect on the whole structure.

I have used four main categories of interactive dimensions – the *socio-economic area, political community, unit of public administration* and *legal system* – to structure the ideal type elements of the cities. In addition to this categorisation, we also need a dynamic perspective, which is given by the (idealised) *historical evolution* process.

I have composed *four ideal types of city governance* to describe the main alternatives for city/agglomeration governance. As far as contemporary cities in Western Europe are concerned just three of these are relevant. Nonetheless, I have come to the conclusion that it is useful to include a fourth category, one that is more American in character: the Private City. Expanding the gallery of ideal types is useful not only for making it easier to compare American and European cities, but also for analysing the emerging local governance systems in Eastern and Central Europe. The latter often resemble the (historical) American cities more than their present-day counterparts in Western Europe.

When drafting these ideal types, I have (over) emphasised, in the spirit of Weber, certain dominant properties of the political and administrative structure of cities. The *Unitary City* exists on just one political level. Both the political and the administrative territories cover the whole urban agglomeration. The *Village City* has just one tier of political decision-making: the municipality. But the political and administrative structure of the agglomeration is split into small units or built up from a group of politically independent villages or small towns. The *Multi-level City* has several tiers of government and a federal type of political and administrative structure.

In all three models of local governance, the political community and its administration play a central role as the main local decision-maker and producer of services. In the *Private City*, the private sector dominates, and the rules of the game are mainly determined by Civil Law.

These ideal types are not parallel in terms of age. Actually, the Village City type of governance is the original model and the historical starting point for all the alternatives of local governance in Europe. The Unitary City is a relative newcomer, developed after World War II as a part of the welfare state. The Multi-level City is an even more modern form of governing agglomerations. The foundation of the Toronto metropolitan government in Canada in 1953 is considered to be the first application of this idea. The first multi-level government systems in European cities appeared in the 1960s (Mennola, 1991).

In the USA the development of local governance, in spite of its European (British) roots, took its own course right from the beginning. The description given by Alexis de Tocqueville of local government and decentralisation in America more than 150 years ago shows amazing similarities with modern US institutions. Private City is an essential part of the American constitutional ideology in general.

We must remember that examples of 'pure' ideal types do not exist anywhere in the world. All real cities are composed of features representing more than one of the models. Certainly the scheme presented here is not the only conceivable typology of cities as political and administrative environments and units. If some other properties are taken as the starting points in characterising cities, the result would be another typology. Openness and flexibility, favouring alternatives and an innovative approach, are other strengths of the ideal type method.

The composition of the four alternative ideal types of city governance are presented in table 4.1.

Using the Ideal Types in Comparative Research

The ideal types presented above are a personal construction of the author. They should be completed, confronted and criticised, of course, as the research and discourse progress. The ideal types are open to continuous dialogue, both in terms of theory as well as empirical research. Actually they are *an initiative to engage in a dialogue.*

The ideal types might be particularly helpful in the following stages of comparative European research and development projects:

- formulating and refining *hypothesis* to be tested by empirical research;
- determining the *special problems* typical of the environment concerned;
- evaluating and improving international *comparability* and *compatibility*;
- developing *recommendations* and model solutions.

The more advanced the characteristics of an ideal type, the more visible the influence of this particular combination of properties on different functions of the city. These hypothetical causalities can be tested by empirical research, assisted by studies of the literature. The properties of the ideal type itself and their verification in the real world may also be worth empirical study.

The relevant features of an ideal type indicate many potential strengths and weaknesses associated with a particular city type. We could foresee that the Village City would have some problems; for instance, when organising large-scale agglomeration services, problems would arise with respect to planning, political decision-making, co-ordination and finance. On the other hand, one could assume that this kind of city is responding quite well, both

Table 4.1   Ideal types of city governance

City governance type	Socio-economic area	Political community	Unit of public administration	Legal system
UNITARY CITY *life circle:* welfare state	• one large unit, created by amalgamations • strict all-round borders • official identity	• cog in the public administration machinery • part of national hierarchies • civil service, expert power, domination • co-ordination by dominance	• unit of service industries • sectored functions • large variety of service activities • large budget and staff, business-like organisation	• national unitary legislation • weak constitutional status • non-statutory protection by professional influence
VILLAGE CITY *life circle:* authority state	• numerous small units, created by evolution • emphasised borderlines in strictly local affairs • local community identity based on traditions	• small-scale autonomy • state domination in large-scale affairs • strong political leadership, weak administration • co-ordination by voluntary co-operation	• real estate owner, producer of neighbourhood services • state and private sector dominance in major services • traditional type authority • non-sector organisation	• national unitary legislation in large-scale activities • weak constitutional status • non-statutory protection by political influence

City governance type	Socio-economic area	Political community	Unit of public administration	Legal system
MULTI-LEVEL CITY *life circle:* coalition state	• several levels of government, created by negotiation and compromises • politically permanent, functionally flexible borderlines • simultaneous multiple identities	• shared autonomy, federal type relations between political levels • local political structures • politician dominance, collective leadership • co-ordination by division of labour	• shared responsibilities based on subsidiarity • dynamics based on financial autonomy • contracted services by private and third sector • non-sectored leadership, functional expertise	• individually defined legal status by contract • limited national regulation powers • federal type legal structures • non-statutory protection by contract principles
PRIVATE CITY *life circle:* enterprise state	• numerous overlapping political and administrative units • functionally changing borderlines • Image-based identity	• private autonomy • delegated powers from below • personal leadership • spontaneous structures co-ordination by market	• mixed types of responsibilities spontaneous, changing tasks • voluntary sector dominance • innovative solutions	• complicated legal structures • private law dominance • non-statutory protection by personal or business influence

practically and mentally, to the citizens' needs regarding many details at the level of their neighbourhood. Administering all the small local details would perhaps be the weak point of the Unitary City. That task might also run into many other problems, manifesting themselves as weak citizen participation, alienation and frustration.

The Multi-level City, adopting federal structures, is simultaneously both large and small. Theoretically, it should be able to deal efficiently with many of the problems typical of both of the previous types. But can we actually find evidence of this problem-solving capacity in the real world? Concerning the Private Cities, social injustice and the democracy gap will most probably be recognised in advance as their biggest problems, undermining their potential advantages in efficiency. But what are the real differences between politician-led and market-driven local development? Is the inner circle of the local political elite actually a more 'democratic' decision-maker than the market, where everybody is able to contribute?

The field of international comparative research is where the ideal type method is most effective, however. An existing gallery of ideal types and comparative analyses of individual cities would give any researcher immediate access to interesting international examples and references. Mistakes caused by innocently imported results from incompatible cultures could be avoided. Differences in urban performance or in the quality and quantity of local services between cities belonging to the same category would raise interesting and probably better defined new research problems. Investigating these would help identify the real causalities. We must always remember that real cities never correspond to their pure ideal types, and that no city is exactly like another.

The potential drawbacks of urban performance in a given city are always related to the interactive political and administrative environment as a whole. The recommendations on how to create more favourable conditions should accordingly take a holistic shape. Propositions concerning renewal in a particular city and its structures could be presented in the form of an ideal type as well. This would mean drawing up a real *Ideal City*, with a combination of properties that would seem most advantageous in promoting urban performance.

The Ideal Type Approach and the real Cities

It is unfortunately beyond my capacity to present all European cities, especially those appearing in this book, as ideal types. My expertise and experiences are not deep enough to adequately. Locating them in categories and motivating the choices would be a major and demanding new research project on its own. Understanding the human nature of a city as a political and administrative environment often presupposes other means to 'experience' it in addition to normal scientific methods.

The European comparative studies I have published for the Helsinki Metropolitan Council, offer a range of observations from which to derive questions concerning some European cities. The comments I make below are more like randomly selected examples than systematic analyses.

There is one relatively pure example of the ideal type of the Unitary City amongst the cases presented in this book. *Oslo*, the capital of Norway, is perhaps an even more authentic representative of this particular city type than its Scandinavian counterparts. That is because geographically it more or less completely coincides with the agglomeration as a whole and has a dual status, being both a region and a municipality. The problem of administering small details and individual services under this heavy structure has already been encountered in Oslo. One way in which these problems have been dealt with is by constructing new neighbourhood councils inside the city organisation to work as local sub-units (Oslo kommune, 1995).

Zurich is the most genuine example of the Village Type amongst the cities represented in this book. Administratively, the Zurich West regeneration project lies within the jurisdiction of the municipality of Zurich, which is the inner-city local authority. Therefore the potential co-ordination problems in large-scale projects can only be observed just indirectly. Such difficulties are obvious, as some other researchers have pointed, although the particular Swiss institution of government by consensus is highly developed to cope with this kind of problem (Arn and Friedrich, 1994; Haldeman, 1997).

Brussels is also closest to being a Village City, although the city was recently institutionalised as a state inside the new federalist constitution of Belgium. The effects of the federal structures at the city level cannot be identified clearly yet, however. Perhaps they will never play an important role in purely local problems. The status of local government in Belgium is relatively weak. It is limited more or less to the traditional tasks of a local civil service. Other forces and institutions, such as the local business community, play an important role. As far as Brussels is concerned, even European institutions can be seen as local power players (Decamp, 1994).

More striking examples of a Private City can be found in prospective EU member states, however. The Estonian capital *Tallinn* and the Polish capital *Warsaw* are typical examples of cities where the business community and private investors dominate local development and politics. Compared with American Private Cities there is a big difference with respect to more or less to everything in Eastern and Central European countries. Many of the 'private' institutions of the latter economies are not basically private but legacies of the former socialist structures. Important local services, for instance energy or water supply, may still belong to the state administration. So the Private Cities of Central and Eastern Europe are unique cases on their own (ILGPS, 1994).

Italian cities are traditionally Village Cities, composed of numerous independent municipalities. Since new legislation was enacted in 1990, it is now possible, to organise a new tier of government at the agglomeration level in nine specified cities as well as in other metropolitan agglomerations. *Milan* and *Turin* belong to this particular group of cities, too. If this kind of metropolitan government is going to be set up, the system will obviously have many of the features of the federalist Multi-level City.

It seems that a problem peculiar to the modern regional and local structures in Italy is the contradiction between the (rather advanced) formal structures and the (relatively modest) financial autonomy of the Italian local and regional government. That might be the case in Milan and Turin, too. Italian cities might thus even constitute a more universal point of reference for assessing the significance of financial autonomy to urban performance.[13]

The French model of agglomeration government dating from 1966, the '*communauté urbaine*', exhibits many features of the multi-level ideal type. One of the ten existing instances is the

Communauté Urbaine de *Lille* (CUDL). The co-ordinating powers of the communauté urbaine over the local authorities are rather strong. That may be why just one new communauté urbaine (Bordeaux) has been set up during the past decade. Other innovations since 1992 include some less integrative alternatives for local co-operation, namely the *'communauté des communes'* and the *'communauté des villes'*, the former having become very popular indeed. Because France has so many municipalities, inter-communal co-operation is of particular importance there (Bernard-Gelabert, 1996).

Federalism is by no means a typical feature of French government. An analysis of the internal structure of any *'communauté urbaine'* reveals many features typical of the federal structures. On the other hand, local structures in federalist Switzerland are not particularly federal at all. The map of 'local federalism' in Europe differs from the traditional view of constitutional federalism.

Working examples of Multi-level Cities in Europe are scarce. Some of the previous institutions have even been abolished (Greater London Council in London, Hovedstadsrådet in Copenhagen). Plans to constitute new forms of metropolitan government have often collapsed in the wake of political controversies, as happened recently in the Netherlands. One of the rare successful examples of recent years is the metropolitan government of *Stuttgart* (Verband Region Stuttgart). The issue of metropolitan government seems to be politically contentious everywhere. It has a very strong influence on the political power structures, both nationally and locally. So strong opposition to any change in the status quo tends to arise from the both directions (Van de Veer, 1998).

It is not difficult to find living examples of links between individual political and administrative qualities of cities and their obvious strengths and weaknesses in urban performance. Oslo, with its unitary governance, put in a conventional metro system many years ago. In contrast, the Zurich agglomeration, with its numerous individual municipalities, is still relying on trams and other more traditional modes of transport. Lille, with its multi-level political and administrative structures, has already created a modern integrated transport system, ranging from international TGV trains and regional traffic to a computer-driven local VAL metro system and pedestrian zones. Meanwhile, the politically and administratively more complicated and turbulent cities of Milan and Turin are just now planning their integrated urban rail systems. Analyses of the political and administrative qualities based on the ideal type method could be very useful in creating circumstances that are more favourable to urban performance with respect to large-scale efficiency and attention to detail.

Conclusions

Comparative research and transfer of knowledge presumes an extensive understanding of Europe and its cities. The purpose of this contribution has been to build a bridge of between the single cases and theories and the surrounding context: the political, administrative and legal environment and its evolution in European cities.

For all of our conclusions, we must recall that we are dealing with the most complex human institutions ever created. Every single city is a product of a unique history. Urban change is continuous and falls largely outside our field of vision. No single discipline alone can produce the necessary comprehensibility and comparability.

Thus, the most important piece of advice we can offer is to find out as much as possible about everything. Political geography, democracy and autonomy, administration and other public activities as well as the legal system constitute the relevant framework for the political and administrative city. All of these features must be observed simultaneously in order to draw the complete picture.

In spite of their many differences, European cities also have much in common. The territorial and political structures have been influenced over the centuries by the same international forces and trends. The qualities of local democracy and autonomy can be evaluated by common European yardsticks of local self-government. Because of the European Single Market and the globalisation of the economy, the production of local services is undergoing similar changes everywhere. Therefore numerous interesting and helpful starting points can be found for our potential classifications and variables.

The confusing diversity of contemporary city life often defies generalisation. It may be useful to seek advice from scholars of the past and derive more timeless starting points for comparison from history. Theoretical construction of ideal types of cities according to the example of Max Weber is one attractive possibility to produce holistic hypothesis that can be tested even by means of empirical research.

I have constructed four basic alternative ideal types of cities reflecting their political and administrative features: the Unitary City, the Village City, the Multi-level City and the Private City. These constructs are meant to serve as initiatives to a dialogue. The goal is to produce a more reliable and comprehensive contextual framework of knowledge for any comparative European research concerning cities.

I am well aware that my theoretical considerations and their application to the real world are going to raise many objections. That is one of the things they are meant to do. Only through extensive dialogue will we get any closer to understanding and implementing the development in our cities.

Notes

1   An impressive account of local self-government as a major element in European democracy is presented by Adolf Gasser (1947) in his classical work.

2   Interesting analyses illuminating the historical roots of the political and administrative institutions in many of the cities observed in the case studies have been presented by, for instance, R.C. van Caenegem (1998), D. Waley (1988) and F. Fleiner (1941). We should not forget to mention Adam Smith here; in 'The Wealth of Nations' he explains why British cities never became such prominent participants in international trade and business as many of their continental counterparts.

3   Oliver Mongin (1995) defines three main 'ages' in the history of cities: first, the period of *road*, when towns and cities were born in terms of transportation and communication; secondly, the period of *factory*, when industrial methods and planning were transferred to city institutions and structures; and thirdly, the period of *chaos, in which* we are living at the moment.

4   According to the arguments presented by J. Douglas Brown (1973), any large company or other business organisation is composed of properties so human that valid generalisations are enormously difficult to make. One single city may include thousands of such organisations plus a multitude of other human actors.

5   According to Paul Feyerabend (1975), even all the conventional scientific methods are questionable in this kind of context, at least as the dominant way of approach, because they neglect innovation and stress verification.

6   Concerning the roots of local government studies, see for instance Haus (1966), Püttner (1981) and von Mutius (1983).

7   E.F. Schumacher (1977) speaks about the four levels of being in the form $(m+x+y+z)$ where m is matter, x is life, y is consciousness and z is self-awareness, the highest and the most important human level that most of the sciences are not able to describe at all.

8   See for instance the works of Samuel Humes and Eileen Martin (1961), Arne F. Leemans (1970), J.J.Hesse and Laurence J. Sharpe (1991) , Edward C. Page (1992), Alain Delcamp (1990, 1994) and Lisbeth Gundlund Jensen (1994).

9   European Charter of Local self-government, Strasbourg 15.10.1985 (Council of Europe / European treaties ETS No. 122).

10   See for instance the article 'The Making of NHS Ltd', *The Economist*, January 21st 1995.

11   Besides studying the work of Max Weber himself, an important way for me to understand his methodology has been to read the articles by the Finnish Weber expert, Tapani Hietaniemi.

12   German economist Werner Sombart and his article 'Der Begriff der Stadt und das Wesen der Städteentwicklung' (1907) probably formed one of the main sources of inspiration for his considerations.

13   See the contribution of Francesco Merloni in Delcamp (1994).

References

Arn, D. and Friedrich, U. (1994), Gemeindeverbindungen in der Agglomeration, *Bericht 37 der NFP "Stadt und Verkehr"*, Zürich.

Batler, R. and Stoker, G. (1991), *Local Government in Europe. Trends and Development*, Macmillan, Basingstoke.

Bernard-Gelabert, M-C. (1996), *L'Intercommunalité*. Paris: Librairie générale de droit et de jurisprudence, E.J.A.

Brown, J. D. (1973), *The Human Nature of Organisations*, AMACOM, New York.

Commission of the European Communities (1992), *Urbanisation and the Functions of Cities in the European Community*. Regional Development Studies report by the European Institute of Urban Affairs. Liverpool John Moores University, Commission of the European Communities, Brussels/Luxembourg.

Council of Europe/CLRAE (1980), *Report on Regional Institutions in Europe*, Council of Europe, Strasbourg.

Council of Europe/CLRAE (1981), *Report on the Principles of Local Self-Government*, Council of Europe, Strasbourg.

Council of Europe/CLRAE (1990), *Types of Financial Control Exercised by Central or Regional Government over Local Government*, Council of Europe, Strasbourg.

Council of Europe/CLRAE (1991), *Financial Self-Government of Local Authorities in Europe, Comparative Study of Local Finance in Six Countries (Belgium, Spain, France, United Kingdom, Italy, Germany)*, Council of Europe, Strasbourg.

Council of Europe/CDLR (1992,) *Borrowing by Local and Regional Authorities*, Council of Europe, Strasbourg.

Council of Europe/CDLR (1993), *Major Cities and their Peripheries. Co-operation and Co-ordinated Management*, Council of Europe, Strasbourg.

Delcamp, A. (1990,) *Les Institutions Locales en Europe*, Que suis-je, Presses Universitaires de France, Paris.

Delcamp, A. (1993), *Regionalisation and decentralisation in certain European countries*, Council of Europe Report CPL/GEN, 93/5, Council of Europe, Strasbourg.

Delcamp, A. (Ed.) (1994), *Les Collectivités Décentralisées de l'Union Européenne*, La documentation Française, Paris.

Feyerabend, P. (1988), *Against Method*, Verso, London.

Fleiner, F. (1941), *Ausgewählte Schriften und Reden*, Polygraphischer Verlag AG, Zürich.

Gasser, A. (1947), *Gemeindefreiheit als Rettung Europas*, Verlag Bücherfreunde, Basel.

Grene, M. (1987), Historical realism and contextual objectivity: a developing perspective in the philosophy of science, in N.J. Nersessian (Ed.) *The Process of Science*, Martinus Nijhoff Publishers, Dordrecht, Boston Lancaster, pp. 69-81.

Haldeman, T. (1997), Die Stadt im Lastenausgleich. Kantonale Programme für kernstädtische Leistungen?, *Zürcher Beiträge zur Politikswissenschaft*, 20, pp. 245-293.

Haus, W. (Ed.) (1966), *Kommunalwissenschaftliche Forschung*, Kohlhammer, Berlin.

Hesse, J.J. and Sharpe, L.J. (1991), *Systems of Local and Regional Government in Western Europe*. Paper prepared for a meeting with the Swedish Association of Local Authorities, Stockholm 18 December 1991.

Humes, S. and Martin, E.M. (1961), *The Structure of Local Governments throughout the World*, Martinus Nijhoff, The Hague.

ILGPS (1994), *Local Government in the CEE and CIS*, Institute for Local Government and Public Service, Budapest (HON).

Jensen, L.G. (1994), Den Kommunale og Regionale Forvaltning i Landene i EU. *CORE arbejdspapirer 3/1994*, Copenhagen.

Leemans, A.F. (1970), *Changing Patterns of Local Government*, International Union of Local Authorities, The Hague.

Mennola, E. (1991), Eurooppalaisen suurtaajaman hallinto, *YTV Pääkaupunkiseudun julkaisusarja 1991/1*, Helsinki.

Mennola, E. (1998a), Eurooppalainen suurkaupunkiseutu 2000. *YTV Pääkaupunkiseudun julkaisusarja 1998/1*, Helsinki.

Mennola, E. (1998b), The European integration and the changing constitutional status of the local and regional authorities. A dynamic perspective, in T. Modeen (Ed.) *The Finnish National Reports to the XVth Congress of the International Academy of Comparative Law*, pp. 175-187, Kauppakaari Oyj/Finnish Lawyers' Publishing, Helsinki.

Mills, C.W. (1980), *The Sociological Imagination*. Harmondsworth: Pelican Books Ltd.

Mongin, O. (1995), *Vers la Troisième Ville*, Hachette, Paris.

Mutius, A. von (Ed.) (1983), *Selbstverwaltung im Staat der Industriegesellschaft*, R.V. Decker's Verlag, Heidelberg.

Oslo Kommune (1995), Strategier for organisering og styring av Oslo kommunes virksomhet, *Bystyremelding Nr. 2* / Oslo kommune, Oslo.

Page, E.C. (1992), *Localism and Centralism in Europe*, Oxford University Press, New York.

Püttner, G. (Ed.) (1981), *Handbuch der Kommunalen Wissenschaft und Praxis*, Springer-Verlag, New York, Berlin, Heidelberg.

Rowat, D.C. (Ed.) (1980), *International Handbook on Local Government Reorganisation*, Aldwych, London.

Schumacher, E.F. (1977), *The Guide for the Perplexed*, ABACUS/Sphere Books Ltd, London.

Searle, John R. (1995), *The Construction of Social Reality*, Penguin Books, Harmondsworth.

Smith, A. (1987), *The Wealth of Nations*. Books I-III, reprinted in Penguin Classics, Penguin Books, Harmondsworth.

Tocqueville, A. de (1956), *Democracy in America*. Specially edited and abridged by R.D. Hefner, Mentor books/Penguin Inc., New York.

Urwin, D. W. (1990), The Politics of Territory and Identity, in SOU 33 (1990), *Urban Challenges*. Report to the Commission on Metropolitan Problems, Statens offentliga utredningar/Statsrådsberedningen, Stockholm.

Van Berg, L., Braun, E. and Meer, J. van der (1997), *The Organising Capacity of Metropolitan Regions as an important factor of Competitiveness*. Contribution to the International Conference on Urban Politics, April 22 1997, conference paper, Helsinki.

Van Caenegem, R.C. (1998), Mediaeval Flanders and the Seeds of Modern Democracy, in J. Pinder (Ed.), *Foundations of Democracy in the European Union*, European Cultural Foundation Huom, Amsterdam.

Van de Veer, J. (1998), Metropolitan Government in Amsterdam and Eindhoven, *Environment and Planning C: Government and Policy*, 16, pp. 25-50.

Wachter, S. (1995), *La ville contre l'état*. Montpellier: Géographiques RECLUS.

Waley, D. (1988) *The Italian City-Republics*, Longman, London.

Weber, M. (1964), *Wirtschaft und Gesellschaft*, Studienausgabe, Kiepenheuer & Witsch, Köln-Berlin.

Weber, M. (1992), *Kaupunki*. The original book "Die Stadt" (1921), Berlin, the Finnish translation and preword by Tapani Hietaniemi, Jyväskylä: Vastapaino.

Zohar, D. and Marshall, I. (1993), *The Quantum Society*, Flamingo, London.

5 Urban Governance by Network Management

WALTER SCHENKEL

Do Policy Networks Matter?

This contribution presents an analytical framework that is specifically applicable to the planning and development processes of many comparable urban development projects (see chapter 1, section 'Governance and Management Perspective'). We start by asking if it is possible and worthwhile to look for new ways to initiate planning processes that give priority to projects integrating living, working and recreation in developing urban areas. By applying the network approach to an analytical task, we can describe the situation in a particular urban area but also see if 'new' instruments make the planning process more efficient and more effective. Given the financial constraints and the established structures for integration and distribution, many towns and cities have tried to at least expand their political scope for activity by increasing their political-administrative power and capacities. This calls for an optimal distribution of legal, financial and organisational resources. But it also involves building forms of co-operation between the various authorities, administrative units and pressure groups. This concerns, on the one hand, democratically legitimised processes and structures, which are firmly anchored in the institutions. On the other hand, it concerns the formation of policy networks for the planning, decision making and implementation phases of programmes and measures. An increasing number of actors, addressee and pressure groups seek to participate in the institutionalised or ad-hoc policy networks in the area of urban planning. Thus, political responsibility is not solely demonstrated by

101

the reduction of budget deficits. It is also shown by the consistent implementation of new forms of co-operation and management.

Urban planning can be project-oriented (emphasising physical and social structures) or process-oriented (focusing on actors, stakeholders and citizens). Either way we have to ask what theoretical support is required. Different disciplines (theory) and sectors (practice) can be guided by *instrumental rationality*; different knowledge values, and interests can be guided by *communicative rationality* (Malbert, 1998). In the field of policy analysis, both building policy networks and designing policy instruments are common ways to analyse how governments select means to address public problems. A policy instrument is defined as a means by which an actor can modify one or more elements of a policy in order to gain a desired result (e.g., prohibition, subsidies, taxation, contractual agreements information). However, choosing a specific strategy does not mean that decision-makers act rationally or calculate the pros and cons of different combinations of measures and instruments. Especially in the context of cross-national comparative analyses, one should look more closely at the cultural and traditional background shaping the decision to adopt one strategy or the other.

New problems call for innovative planning, which includes experimentation, limit extension, and a new form of organisation. Contractual agreements, for example, are a 'new' form with a crucial precondition: there must be a positive attitude expressed by the target groups. Once the development problem has taken centre stage, a new model of management requires government agencies that can bring about processes of change. This involves bridging the gap between long-envisioned aspirations and short-run opportunity (Strauch 1996). The government has to mobilise multi-actor systems having following characteristics:

I   no single actor can attain the defined objectives and there is a *market of decisions;*
II objectives are put into a broader development perspective, so that *win-win situations* are more likely;
III visions develop in a *package deal* that, on balance, leads over time to greater gains than losses for each societal actor involved.

The increasing popularity of voluntary agreements may be seen as an expression of the state's shifting role in many policy fields. On the other hand, one should not forget that old consensus-oriented democracies such as Switzerland have deeply rooted traditions of negotiation. These come into play early in the policy-making stage when the parties can focus on creating regulatory instruments. However, urban planning projects can offer new opportunities to open a so-called *policy window*; once the champions of a particular solution sense some movement, they leap into the arena to promote their ideas (Kingdon, 1995). It is assumed that urban planning activities, decision-making and implementation are no longer hierarchically structured processes. Therefore dynamic research dimensions may best be explained according to some elements of the *policy network approach* (van Waarden, 1992; Scott, 1991). Networks consist of a variety of actors, conflicting interests and highly dynamic features such as an arena to make choices among problem-oriented, combined sets of alternatives, divided power, unclear information and a broad variety of goals. The network approach is not only an analytical tool but includes network management and network building. Why does it make sense to focus on actors rather than physical urban structures? (Knoke and Kuklinski, 1991):

- *actors prefer to interact with those who appear to be deeply involved in the decision-making process.* The more the actors who are particularly affected by the problem take part in the networks, the more likely that specific knowledge can be used to work out better solutions;
- *exchange relations are basically a series of transactions that become more regular and profound if the network remains active over a long period.* Complex problems require not only limited adjustments of institutions but also profound reorientation of the political system;
- *actors' positions have to be considered when looking at the individual view of problems, behavioural aspects, and the network output in general.* Challenges posed by complex problems may alter the political discourse and help promote institutional reforms. To mount a strong repines, there must be a supportive actor in a

central position which enables him to transform new ideas into network relationships.

The application of the network approach can contribute to a better understanding of the determinants shaping the *discourse* within urban planning networks and the choice of policy instruments. Furthermore, putting the approach into practice can help clarify the preconditions and functions of the *learning processes* occurring within policy cycles in networks that produce planning decisions. Last but not least, the network approach is assumed to serve as a good research tool to explain the very specific problems of our case studies on urban planning:

- we have well-defined urban areas;
- we have some loose networks, with a limited number of actors, from which we can start our data collection concerning actors and relationships (according to the snowball principle);
- the network approach is open enough to analyse dynamic dimensions such as changing actor constellations, changing belief systems and changing instrumental priorities as well as stable dimensions such as the political, spatial and societal setting in which the planning process is embedded;
- the network approach is not only applicable under controlled scientific conditions but also under the shifting conditions of an ongoing and incomplete planning process.

In the context of European integration and globalisation, city governments, urban planners, and developers pose a probing question: *Can urban management make cities more competitive?* Bramezza (1996) has identified six dimensions which affect the position of cities in the European urban hierarchy:

I   globalisation and formation of continental blocks;
II the shift to informational economy;
III impacts of transport technology;
IV impacts of information technology;
V   impacts of demographic and social change;
VI urban promotion and boosterism.

The global economy produces 'global' cities. Economic diversification was seen as one of the key factors leading to the success of particular cities such as Hamburg, Rotterdam and Dortmund. The message is that throughout Europe, increasing competition and the priority given to economic objectives has led to a fragmentation of the planning process and a greater involvement of the private sector. Therefore, cities have to develop visions and have to translate them into strategies and goals. This process only functions within policy networks and by taking political and spatial-economic conditions into account (Bramezza, 1996).

From Network Analysis to Network Management

Ignoring the existence of structured actor networks for the moment, the processes of information exchange, negotiation and decision-making may turn out as follows (Scharpf, 1993):

- the actors' positions remain *conflictive*. Top-down decision-making or (direct) democratic procedures have to seek solutions. Actors invest resources in political campaigns and lobbying. Law plays an absolutely dominating role. *'Best solution' means that the most powerful actor will be the winner;*
- the positions start to change and actors reach *consensus*. These solutions are non-binding and have to be approved by political authorities or through (direct) democratic procedures. Actors jointly invest resources in pushing their priorities onto the political agenda and in finding majority backing in the broader public. Law plays a dominating role, especially at the beginning and at the end of the process. *The 'best solution' will be the most feasible on;*
- actors agree on given framework conditions and a set of alternative measures that they jointly elaborate. They are willing to draw up *covenants* to continue a non-hierarchical solution-seeking process. Actors invest resources in network management and network building. Strict law plays no longer a dominating role; skeleton legislation leaves room for flexible and cross-cutting solutions. *The 'best solution' will be an accepted and institutionalised arrangement to create optimal win-win solutions.*

The research proposals should not compete with research ideas coming from urban planning, sociology, and economic theories but contribute to the COST-CIVITAS programme from a political science viewpoint. It is thus objective of empirical research to discover how to create institutions for an efficient and effective urban planning policy. Two aspects of policy networks should be emphasised. First, they represent *new forms of collaboration* between the state and society. Second, networks redefine the individual actors' scope for action. Potential power and influence can be derived from *access to* and *individual positions within a network*, and not only from the actors' financial and political resources. Access and position are basic network ideas. In that context, authorities are not treated as separate entities outside the set of policy actors. Rather, they are seen in terms of specific roles played by a subset of actors, namely public actors (Knoke and Kuklinski, 1991). Our research questions must distinguish clearly between four objects of observation (Kenis and Schneider, 1991; Scott, 1991):

- *who are the actors?* Areas of interest are: legal, financial, political, and knowledge-based resources; specific interests; actors' background organisation; actors' network position (core, periphery, gatekeeper etc.);
- *how can we describe the quality of actors' relationships?* Areas of interest are: (in-)formal, (non-)hierarchical, (non-)directed and (un-)equal relationships; frequency of contacts; relationship typology such as information exchange, bargaining, negotiation, co-ordination, collaboration, decision-making and implementation;
- *how can we describe the network as a whole?* Areas of interest are: network density, structure and function; political, spatial and societal context; legal, financial, organisational, and political resources of the network; conflicts and coalitions within the network; supportive actor with new ideas;
- *where are the interdependencies between the network and its environment?* Areas of interest are: external framework conditions and neighbouring planning processes; political, economic and social situation in general; local, regional, national and international developments; relevance of the network's output for its environment; control over activities carried out in

the name of the network and its influence on the network's environment.

City governance is assumed to reflect a new understanding of decision-making and steering. The general picture - at least in democratic political systems - is one of increasing differentiation of responsibilities for managing *'new' complex problems*. Here it is argued that strict law enforcement and 'artificial' separation of public and private space should be eclipsed by flexible collaboration forms. Discursive interaction can create new meanings and new identities in urban areas. In that way it plays a key role in processes of political change, when 'economic growth coalitions' overlap with the interests of inhabitants. Policy making – or, in our context, urban planning – is not simply an abstract process of resource exchange. Actors involved in the process justify their policy decisions by using arguments derived from ideas, beliefs and values already present and established. Successful consensus-seeking is not only dependent on having the 'right facts' at hand.

Specific questions are focused on changing network dimensions such as the shift from *capacity-using* to *capacity-giving* use of power. Capacity-giving networks are assumed to give high priority to knowledge-based power rather than to financial and political power. In other words: argumentative discourse and network-building activities take precedence over political struggle, dispute and confrontation. Interaction, open information channels, and neutral process mediators may lead to learning processes among all actors. 'Old' ideologies erode and the actors' belief systems shift towards a common vision, including some unconventional policy experiments. Furthermore, uncertainty, doubt and value-oriented argumentation must be considered as crucial elements of open non-hierarchical networks. Participants must be convinced that talking about uncertainties and experiments is the most efficient way to create win-win situations (Cameron and Wade-Gery, 1995). In the newer literature, *reflexive communicative steering processes* are often seen as the only alternative that offers long-term and durable solutions to compensate the state's lack of capacities and evidence of action. However, our core research question must be broader, covering the problems of strategic control, of sufficient information, of democratic guidance, of common language and of sufficient motivation (Friend and Jessop, 1969): *What kind of steering instruments do we have at our disposal to achieve more efficient and more effective economic, political and social processes in order to deal with urban development problems?*

Policy networks are more or less stable patterns of social relationships between interdependent actors. These relationships take shape around policy problems and programmes. They are continually being formed, reproduced and changed by 'games' played by these actors. A game is an ongoing, sequential chain of (strategic) actions between different 'players' governed by formal and informal rules. The network approach must not be understood as a social science theory but as an *analytical toolbox* combining and testing different theoretical concepts and assumptions (Klijn, 1996; Kenis and Schneider, 1991). Furthermore, the network approach implies some very *practical meanings*. Besides serving as a methodological framework for qualitative and quantitative empirical research, it offers ideas on concepts how to manage and restructure existing actor networks and how to build up new ones. Hence a stepwise application can be considered:

- *network analysis* means observing the actors' participation in a network, analysing their resources and relationships, describing their interests, and distinguishing between network inputs and outputs. At a practical level, a network analysis can contribute to a better understanding of conditions under which actors may agree on consensual targets. Potential power and influence can be derived from individual positions within the network. If research results reveal shortcomings in the network and planning process, we may enter the phase of network management. In other words, we must discuss potential improvements;

- *network management* means bringing in new ideas, reflecting on the network's features in the light of research results and evaluations, stimulating discourse and facilitating learning processes. Here the actors' perspectives and preferences are seen as variable rather than constant. Part of the management challenge is to recognise circumstances in which apparent deadlock can be moved towards a more favourable outcome. If the participants reach some agreement on how to review and revise the network itself and the planning process, we may enter the phase of network building, *i.e.* we must consider new participants;

- *network building* means bringing in new actors, changing actors' positions, restructuring actors' relationships, changing interaction rules, opening new communication channels, and redefining the network's function in its environment. That is we must ask how to change the network's constellation in order to change policy as such. Here co-operative action is aimed not only at resolving problems but also at changing the context of the network so that new perspectives may become apparent. Managing the context means taking other policy fields and other actors – sometimes even new conflicts – into consideration, not least to improve the chances of eventual win-win solutions. Network building is a process of gathering information about the behaviour needed for long-term impacts.

Network management is quite distinct from the ordinary tasks of management (*i.e.* strategic planning, structuring and leading the organisation). Empirical research suggests that the restructuring of urban planning processes takes place in well-organised networks. The main actors are city authorities and private actors, whereas tenant organisations and project developers find themselves at the periphery of the network. On the other hand, it is assumed that actors lacking access to direct collective bargaining will strategically invest more resources in opposing or watering down the policy proposals rather than attempt to influence the kind of instruments selected in the first place (Hanf and Koppen, 1993). Radical changes in formulating problems and seeking solutions can take place only if existing rules lose their regulatory power. This means, for instance, that network management and – at a later stage – network building must adopt an open-minded and 'border-crossing' stance. Then, strict law enforcement and power will no longer predominant, and the scope for possible solutions will be broader. Planning processes must go beyond the traditional dualism between private property and public space. Only open networks leave room for unconventional ideas and an integration of opposing interests, often brought in by creative people. An *early entry into the game* – sometimes a re-activation rather than an entry – can lead to significant agenda changes and better implementation records. This is in line with the hypothesis that generalists, co-ordinating agencies and communicative mediators should be given a strong position within urban planning networks. Highly specialised actors who have not been subjected to learning processes tend towards confrontation, calling for law enforcement, rather than towards making compromises on their 'technical' findings. Network building may result in restructuring regulatory space and setting preconditions for a particular kind of co-operative self-regulation. The legal power of the state and strict law enforcement become unattractive because communicative steering is seen as more efficient. Planning objectives can be put into a *broader development perspective*. On the one hand, the network manager can adopt an active strategy, i.e. guiding activities from the top down. On the other hand, he can take a wait-and-see approach. This means limiting himself to supporting initiatives from the bottom up, i.e. from local authorities, private parties or associations, and helping initiatives in similar areas to network together.

Research Model: Framework and Network

Figure 5.1 summarises the research process and the variables contained in hypotheses derived from political science theories in a governance and management perspective. The research procedure may involve three steps. The first is creating the so-called *story line* of the urban planning project by studying the literature, documents and legislation and by conducting structured interviews. The second step is analysing *networks* in quantitative and qualitative terms, including categories of 'belief systems', the number and characteristics of actors, the quality of relationships as well as network density and structures (through structured interviews and questionnaires). Step three consists of making *network-management guidelines* and *network-building proposals*. In summary, the comparison between different projects and countries should give insight into the changing role of governmental authorities, clarify the role of new strategies, resolve the problem of legal and scientific uncertainties, relate network qualities to learning processes, chart

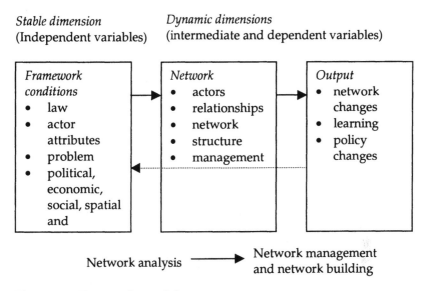

Figure 5.1 is described by the following text layout:

Stable dimension
(Independent variables)

Dynamic dimensions
(intermediate and dependent variables)

Framework conditions	*Network*	*Output*
• law	• actors	• network changes
• actor attributes	• relationships	• learning
• problem	• network structure	• policy changes
• political, economic, social, spatial and	• management	

Network analysis ⟶ Network management and network building

Figure 5.1 Research model

the degree of social and environmental incorporation, and guide the choice among short-term and long-term options. By taking the network approach into consideration, we gain the opportunity to bridge the gap between theory and practice. Concepts of argumentation, which are based on different national practices and traditions and may have an impact on policy outcomes, are assumed to serve as the methodological bridge. Crossing that bridge may lead to a meaningful system of comparative categories.

References

Bramezza, I. (1996), *The competitiveness of the European city and the role of urban management in improving the city's performance*, Tinbergen Institute, Rotterdam.

Cameron, J. and Wade-Gery, W. (1995), Addressing Uncertainty, in B. Dente (Ed.), *Environmental Policy in Search of New Instruments*, Kluwer, Dordrecht, pp. 95-142.

Friend, J.K. and Jessop, W.N. (1969), *Local Government and Strategic Choice*, Tavistock, London.

Hanf, K. and Koppen, I. (1993), *Alternative Decision-Making Techniques for Conflict Resolution. Environmental Mediation in the Netherlands*, EUR, Rotterdam.

Kenis, P. and Schneider, V. (1991), Policy Networks and Policy Analysis: Scrutinising a New Analytical Toolbox, in B. Marin and R. Mayntz (Ed.), *Policy Networks. Empirical Evidence and Theoretical Considerations*, Campus, Frankfurt a.M. and Boulder, CO., pp. 25-62.

Kingdon, J. W. (1995), *Agendas, Alternatives, and Public Policies*, Harper Collins, New York.

Klijn, E. (1996), *Regels en sturing in netwerken. De invloed van netwerkregels op de herstructurering van naoorlogse wijken*, Rotterdam, Eburon.

Knoke, D. and Kuklinski, J. H. (1991), Network Analysis: Basic Concepts, in G. Thompson (Ed.), *Markets, Hierarchies and Networks: The Co-ordination of Social Life*, Sage, Beverly Hills, pp. 173-182.

Malbert, B. (1998), *Urban Planning Participation. Linking Practice and Theory*, Dissertation at Chalmers University of Technology, Gothenburg.

O'Toole, L. J. (1997), Managing Implementation Processes in Networks, in W. Kickert (Ed.), *Managing Complex Networks. Strategies for the Public Sector*, Sage, London, pp. 137-151.

Scharpf, F. W. (1993), Positive und negative Koordination in Verhandlungssystemen, in A. Héritier (Ed.), *Policy-Analyse. Kritik und Neuorientierung* (PVS Sonderheft 24/1993), Westdeutscher Verlag, Opladen, pp. 57-83.

Scott, J. (1991), *Social Network Analysis. A Handbook*, Sage, London/Newbury Park/New Delhi.

Strauch, V. (1996), Zur Entwicklung des Stadtforums, in H. Kleger, A. Fiedler and H. Kuhle (Eds.), *Vom Stadtforum zum Forum der Stadt*, Fakultas, Berlin, pp. 85-104.

Waarden, F. van (1992), Dimensions and Types of Policy Networks, *European Journal of Political Research, 21*, Special Issue: Policy Networks, pp. 29-52.

6 Time, Mobility and Urban Governance: the Case of the Metropolitan Area of Milan

MARIO BOFFI AND GIAMPAOLO NUVOLATI

Introduction

According to urban sociologists, the contemporary large city is better defined as a *third-generation metropolis*. It is the result of a deep transformation of the ways of living, working and consuming performed by different population groups: residents, commuters, city users and businessmen (Martinotti, 1993). This framework produces a great variety of flows of people, which leads to competition for the use of space. The growing complexity of the urban system generates a multilayered timing based on the fragmentation and combination of labour, consumption and social relation's activities.

Time presents a private as well as a public configuration and regulation. As a private dimension, time is linked to the distribution of domestic/familial resources, roles, physical needs, habits and capacities. At the same time, it concerns the organisation of the community – mainly in terms of production and consumption patterns – in which individuals live and work (Bastian, 1994; Hamermesh, 1996; Dhondt, 1998; Gauvin and Jacot, 1999). Therefore, time may be considered as a result of mutual adaptation between collective and personal perspectives.

Time is a limited resource. It creates conflicts but also co-operative pacts between individuals and between collective actors, leading ultimately to a more rational production system as well as to the provision of services. Its regulation is based on formal negotiation between social and economic groups (managers, workers, shopkeepers, consumers, public administrators). Though their interests and needs differ, they share common strategies to improve their level of social satisfaction. Co-operative pacts are, of course, highly diffuse. They thrive in the informal relationships that exist between members of the same group (small or large, they could take the form of a family, neighbourhood, association, and so on) and can contribute strongly to the quality of life of the members themselves in terms of improved availability and better distribution of time. Nowadays, there is a profusion of co-operative pacts in the public context. They take the form of debates and agreements between collective actors focusing on time management.

In other words, uses of time are relevant at various levels. Schedules are based on different models of socio-economic structures as well on informal private practices. During the last three decades, time became one of the most important issues for analysing living conditions and life styles of the population as well as for promoting urban policy interventions. This chapter examines the relationship between time use and urban governance. The issues are discussed from a theoretical perspective (sections 'Time uses and Life Styles', 'Urban Time' and 'Typologies of Time Use') and from an empirical one, highlighting the case of the metropolitan area of Milan, where time regulations were pioneered in urban policies.

Time Uses and Life Styles

The origin of empirical studies on time can be found in the tradition of *time-budget analysis*. More specifically, it can be found in research to evaluate people's sense of *well-being* according to their use of time. Studies on the subject started with the work of Szalai (1972) in the 1970s and continued over the years (Juster and Stafford, 1985) and are still appearing (Gershuny, 1995). These studies investigate the availability and distribution of human time and the set of factors that determine the effectiveness of time use and generate individual well-being. According to that tradition, time opportunities and constraints are crucial, in determining the quantity and quality of social relationships, consumption patterns and labour activities.

Therefore, they constitute important inputs for a new definition of social and gender disparities. Any new definition must take these variables into account, as they reflect not only the availability of money and material resources, but also the integration of formal and *informal* economic issues related to time use.

The interest in time increased recently, mainly in connection with the analysis of spatial mobility and relational networks, as well as with the analysis of the role that technological devices play in people's private and working life. Alternative paths in the use of time, space and information technology are now part of a process of a private *trade-off*. People weigh the relative advantages of commuting practices, level of information and social participation, family organisation and, at a more general level, life styles. In particular, telecommunication devices – fax, mobile phone and computer – are responsible for the rapid evolution of private and public life and of the systems of production and distribution.

Since the 1980s, several economic investigations have studied the evolution of the labour market and the urban system. That evolution was studied in connection with the process of decentralisation of the activities and linked with technological innovation (Camagni et al., 1984; de Lavergne, 1985). More recent sociological studies emphasise the role played by people's level of confidence and competence in performing modern *self-service* practices necessary to query the information systems or to access a service. People have been comfortable with the use of cards for shopping and banking, with the idea of home robots, and with working telecommunication devices (Gershuny, 1993; Paolucci, 1993).

New forms of inequality and segregation are emerging. These are not directly linked to socio-economic classes. Rather, they are tied to different patterns and opportunities in the use of time and modern technologies. Consequently, the established types of social stratification, integration and exclusion have to be revised. New relevant questions are emerging. How can technological competence in everyday devices transform the time-space relationships? Would they reduce or reinforce social and economic disparities?

Urban Time

Once we have ascertained the importance of the time dimension in determining life styles and public policy, we have to investigate if the *urban* dimension matters in time use and governance (Mückenberg, 1998). Cities are places where the modern complexity takes the most evident and controversial in terms of opportunities and *externalities*, conflicts and cohesion/participation. Despite the spread of cities at the metropolitan and regional level, the single administrative units are developing independent urban policies to reduce the complexity and to solve the problems of citizenship. Time policies are therefore formulated and implemented primarily at the local level, although the appropriate framework is the city-region or the metropolitan area. On the other hand, the autonomy and capacity of the local authorities are crucial resources for matching the time needs of the population and improving the quality of urban life. The process of bargaining between different actors and organisations, private and public, in the cities calls for the constitution of permanent arenas where emerging problems can be confronted and resolved. In view of the transformation of the cities and their governance, there is an urgent need to draw up the theoretical and empirical typologies of the use of time and space based on sociological attributes. Meanwhile, it is also critical to identify and test time policies to see if they serve the needs and life styles of different population groups.

Typologies of Time Use

Modern society is characterised by a strong fragmentation of time. Metropolitan life in particular is not based on standardised and strictly sequential times – mainly work/non-work times – but is grounded in a set of more heterogeneous and flexible times to be combined in a *puzzle* by each individual (Treu, 1986; Bosch, 1986).

According to Manacorda (1996), time distribution can be related to specific activities as well as to more general categories:

- *necessary time* (to sleep, to eat, etc.);
- *compulsory time* (to work, to study, etc.);

- *bounded time* (to use services);
- *free time* (leisure, hobby, social relations).

For each of these categories, technological innovation can partially or totally substitute for previous specific activities. Therefore, new technology can save time or make the available time more flexible. For example, in the case of *compulsory time* activities, the high degree of complexity of the operations, the high level of socialisation required of the actors, and the collective perspective of the actions themselves determine the possibility of only a partial substitution (teleworking or distance learning). In the case of *bounded time*, the individual use of the services, the low level of socialisation, and the relative simplicity of the operations can favour the complete substitution of specific activities (telematic windows, home banking, and so on) and save a large amount of time.

According to these general dimensions we can draft a typology of life styles. This does not require adopting a normative model (technological innovation is good or bad). Rather, it means defining a set of needs, habits and capabilities of different population groups in exploring and exploiting alternative resources and opportunities. Such an approach overcomes the drawbacks of neo-contractualism (minimum set of resources guaranteed) as well as those of the classical utilitarian approach (satisfaction of the people) in defining well-being. Furthermore, the proposed approach is more appropriate to evaluate what Sen (Nussbaum and Sen, 1993) calls *capabilities*. That term denotes the ability of an individual to perform specific activities according to a specific set of values. In that sense, the quality of life corresponds to the level of freedom and knowledge of the people in selecting and performing practices (*functionings*) that can contribute to the complete self-realisation of the individual.

In this framework, time need and technological attitude can be crucial variables, being indicators of alternative life styles and desiderata. Of course, different patterns are conceivable. For instance, elderly people may not be interested in having more time or in utilising technological devices. However, managers or workers in particular sectors could be strongly oriented to the use of technological devices in order to save time without reducing their personal productivity, and to use the time saved for leisure.

Urban policies that define different actions in time policy regarding different population clusters are emerging. As a matter of fact, a strict standardisation of the procedures related to time planning is difficult. It does not necessarily improve the quality of life. Instead, a better way to deal with urban complexity would be through different policies targeted to a highly segmented society. This is a way to avoid cultural and social exclusion.

The evolution of empirical research in the time field followed the course of the theoretical development mentioned above. From time-budget analysis to study the availability of time for different activities (objective approach) and different population groups, the focus has rapidly shifted toward studies aimed to verify the level of satisfaction of the people for the distribution of time (subjective approach). The two Social Surveys in Lombardia (IReR, 1991 and 1995) are examples of such a trend. They combine questions about time distribution by type of activity with questions regarding people's opinions on the availability of free time and what can be done with it. Recent research seeks to combine not only objective and subjective data on time use, but also information regarding spatial mobility, technological attitudes and other sociological dimensions. An interesting typology has been designed for a study supported by Centro Studi San Salvador (1996) on time use in Italy. The resulting clusters were placed along a continuum. At one end there was 'too much time available'. At the other end, there was 'no time at all'. In between was the 'right time'. Each cluster shows a level of social exclusion/integration, a measure of attitude toward using technological tools for different activities, and a time-budget profile.

Time-Space Mobility as the Emerging Issue

Profound changes in the structure of contemporary cities have drawn attention to the problem of social, economic and political governance of new metropolitan complexes. Traditional municipal policies and institutions are not up to the task of governing these new entities. New phenomena related to the transformation of cities are arising. In particular, spatial mobility is emerging as one of the major features of urban areas. Indeed, spatial mobility seems to be the basic trait underlying the way of life of a large share of the urban population. Mobility and time use practices are closely linked. This pair of factors is crucial in reshaping living conditions. But they also show a mix of patterns related to different types of populations living, working and shopping in the city: residents, commuters, city-users and businessmen. These populations engage in co-operative as well as conflicting relationships, in sharing the limited resources that are concentrated in urban areas.

Nowadays, the mobility of people within metropolitan areas has composite features. These are mainly related to the rejection of the classical and simple bi-directional paths, *house-office or house-school*. Now, mobility is more oriented toward a sequential combination of fragmented activities (work, travel, use of services, personal relations).

In fact, new forms of spatial behaviour are emerging in the metropolitan areas. Produced by different population groups, this behaviour is based:

- on a rational use of available space and time;
- on the actual organisation of labour in the production system;
- on the efficiency and distribution of the transport and service networks.

As a result of the change in people's time-space practices, innovative sociological models or patterns of mobility are emerging.

According to recent figures,[1] more than 31 million Italians commute everyday to work or to study. There was an increment of 200,000 commuters between 1995 and 1998. Most of the mobility paths (59.6%) lie within the place of residence or run between two communes in the same province (27.8%). In 1998, mobility not related to work or study activities but to interpersonal relations, shopping and cultural events involved 13 million people per trimester. Finally, in 1998 almost 14 million business trips and more than 62 million overnight hotel stays for the same reason have been registered.

With the growing complexity of urban society, there is a corresponding growth in the responsibilities of local governments to manage such complexity. It is therefore necessary to explore and monitor, even from a quantitative point of view, the transformations occurring in different sectors of modern life. New life styles are emerging, corresponding to the way individuals deal with mobility problems. Statistics on the motivation, duration and frequency of the mobility are crucial to the development of planning strategies.

Whereas data on population flows and transport systems turn out to be widely available, data on spatial mobility and time use are quite rare. Moreover, the relationships between a person's work schedule, use of services, and satisfaction with the opening hours of the services themselves have been surveyed infrequently.

On the other hand, the common statistics on use of space and mobility take for granted that the focus must be put on the inhabitants, or perhaps more accurately, on the overnight population. We know very little from the official statistics about the daytime population, which can be considered the real city. Nor do we have any systematic knowledge about the time-space profile of the people who use the city. On the one hand, a large share of all urban problems are related to mobility – traffic, air pollution, competitive use of space. On the other hand, there is insufficient knowledge about the actors and their performance on the urban scene.

The Metropolitan Area of Milan as a Test Case of the Mobility Models

The metropolitan area of Milan has been defined differently by geographers, economists and sociologists. A commonly accepted definition based on structural and functional criteria – as proposed by Svimez (Cafiero and Busca, 1970) according to the Standard Metropolitan Statistical Area – roughly recognises the extension of the metropolitan area as being 150 km wide, 80 km long, geographically centred between Milan and Bergamo and functionally centred on Milan, as shown in Figure 6.1.

The area counts around seven million people. The core of the area lies in Milan, which has 3.7 million inhabitants in the province. Mobility indicators in this area, which consists of 188 communes, show that at least 2.1 million people, who represent 56% of the population, move around to work or study (Table 6.1). A large

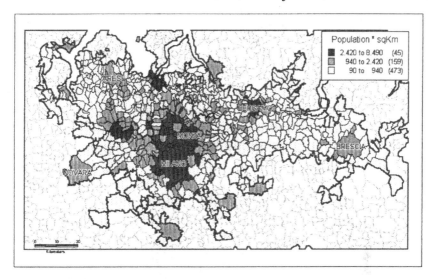

Source: Istat, Demographic data, 1997

Figure 6.1 Metropolitan area of Milan, population per square km, 1997

share, from one-third to one-fourth, of those 2.1 million travel to another commune and a significant proportion of the population spend more than half an hour on a one-way trip.

Those figures underestimate the true mobility. They are not updated, nor do they account for different relevant mobility sources

Table 6.1 Province of Milan, mobility indicators

	Value *1000	% on population
Population	3,732.0	
Workers and students daily moving:		
Total	2,110.7	56.6
Moving inside the city where they live	1,210.1	32.4
Moving outward the cities *	900.6	24.1
Moving inward the cities *	1,036.1	27.8
Moving trip > 30'	492.9	13.2

(*) Included persons moving in and out the province

Source: Istat, Census 1991

or actors, such as the city users. At least they give some idea of the relevance of mobility to the functional model of the metropolitan area.

Two other views of mobility help depict the impact on urban life. The first concerns the extent of the daily flow of population in the administrative units, which shows variations from night-time to day-time population of about 40% on average (see Figure 6.2). The second concerns the geographical spread of mobility, not limited to the core city of Milan, but covering most of the metropolitan area (see Figure 6.3).

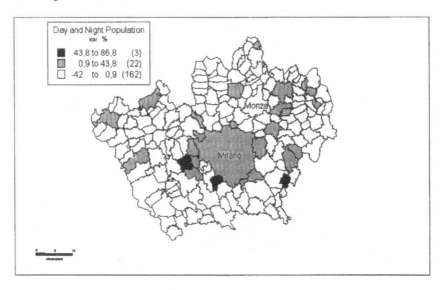

Source: Istat, Census 1991

Figure 6.2 Province of Milan, day and night population, percentage of variation

Daily Mobility Patterns

A relevant aspect of mobility refers to the different patterns adopted by groups of city users or inhabitants. In fact, in order to adopt policies founded on time, an analytical knowledge of the time-space map is needed.

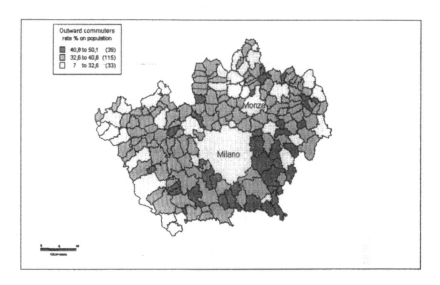

Source: Istat, Census 1991

Figure 6.3 Province of Milan, outbound commuters, percentage of population

A preliminary analysis of urban calendars from a sociological point of view shows a great variety of daily mobility patterns. These patterns vary according to the social group and depending on several factors. For instance, if we trace the calendar of daily departures from home, either in the central city of Milan or in its *hinterland* (Table 6.2 and 6.3), we discover the differential impact of localisation on the mobility pattern of workers and students. Workers show few differences either in terms of the time of leaving

Table 6.2 Mobility patterns: time of leaving home in hours and minutes

Residents in:	Workers	Students	Pupils	Total
Milan	7.51	7.55	7.57	7.52
Hinterland	7.41	7.33	8.01	7.42
Metrop. Area	7.45	7.41	7.59	7.47

Source: Istat, Indagine multiscopo sulle famiglie, 1996-1998

home or in terms of length of the trip to work. In contrast, students aged over 14 display very different time patterns in the central city and in the *hinterland*. The calendar of pupils under age 14 seems to be hardly affected by the territorial context.

Table 6.3 Mobility patterns: length of trip to work or school in minutes

Residents in:	Workers	Students	Pupils	Total
Milan	26	29	12	25
Hinterland	27	49	9	28
Metrop. Area	27	40	10	27

Source: Istat, Indagine multiscopo sulle famiglie, 1996-1998

Further sociological analysis of the calendar demonstrates the relevance of professional status and gender as determinants of the mobility patterns (Table 6.4). Blue-collars workers and employees generally leave home earlier than managers or executives. While women and men of the latter group differ very little, women employees show a clearly different pattern. Those figures suggest that the parameters influencing the patterns of mobility refer to several factors. One is the spatial distribution of the functions and the social stratification in the urban area. Another is the economic organisation, part of which is post-fordist, while much is still fordist.

Table 6.4 Mobility patterns: time of leaving home and trip length in hours and minutes

Profession	Time of homeleaving			Trip's length
	Male	Female	Male	Female
Manager	7:57	7:59	7:57	0:31
Executive	7:59	8:08	8:02	0:27
Clerk	7:43	7:59	7:52	0:30
Blue collar	7:15	7:46	7:25	0:24
Employee	7:39	8:30	7:50	0:19
Professional	8:23	8:40	8:27	0:26
Self-employed	7:28	7:37	7:30	0:19

Source: Istat, Indagine multiscopo sulle famiglie, 1996-1998

Family roles, are also a factor, differentiating women's patterns according to the sociological context.

A different approach is useful for analysing the mobility and the time use. That approach considers the two main types of working schedules commonly found in Italy. One is based on the availability of a break for lunch; the latter has no lunch break.

In the Province of Milan[2] workers having a time schedule with lunch break represent 91.3% of the active male population and 71.8% of the active female population. The average travel time to go to work is about 35 minutes for people living and working in Milan, 29 minutes for people living and working in the hinterland, and more than 50 minutes for commuters between the hinterland and Milan (Table 6.5).

Table 6.5 Trip to work, one-way journey (%)

Living	Working	0-20	20-40	40-60	>60 min.	Mean *
Milan	Milan	29.4	35.9	26.6	8.1	35
Hinterland	Hinterland	45.5	29.6	17.0	7.8	29
Hinterland	Milan	3.7	25.1	40.6	30.6	53
Milan	Hinterland	3.1	31.1	52.7	13.1	51

(*) Average time on two ways: go and return (minutes)

Source: Dept. of Sociology and Social Research (Univ. of Milan-Bicocca), 1998

People having a schedule with a lunch break depart from home at 7:35, arrive at work at 8:11, and pause from 12:22 till 13:33. They quit work at 17:40 and arrive home at 18:16. The 'unsplit schedule' workers have, of course, no formal break. They generally quit work at 13:15 and are back home at around 13:48 (Table 6.6).

Timing of Services

A second type of indicators of the urban calendar concerns the user's timing and the level of satisfaction for the timing of some very common services. The national survey offers some insight into the

Table 6.6 **Daily timetable of workers**

	Schedule with lunchbreak	Unsplit schedule
Going out	7:35	7:43
Arrival at work	8:11	8:15
Start the pause	12.22	
End of the pause	13:33	
End of work	17:40	13:15
Arrival at home	18:16	13:48

Source: Dept. of Sociology and Social Research (Univ. of Milan-Bicocca), 1998

level of satisfaction for different groups of people. The main finding that might be drawn from the figures in Table 6.7, is the differential dissatisfaction with the timing of three services. The table indicates the complexity of the calendar of some groups of people, mainly working people, and the relevance of the timing of services as a component of the quality of the life.

Table 6.7 **Dissatisfaction with the timing of services, metropolitan area of Milan, percentage of total**

Condition	Registry Office	Social Service	Post Office
Working people	35.6	34.1	26.6
Employee	37.1	35.9	28.3
Blue collar	29.8	28.3	18.0
Professional	45.6	39.9	29.4
Housewife	12.2	17.4	14.1
Student	28.7	28.6	27.3
Retired	8.0	14.4	12.1

Source: Istat, Indagine multiscopo sulle famiglie, 1996-1998

Another indicator of the relevance of timing is the very thin calendar slicing found among working people. In fact, during the lunch break, many of the workers use public services (Table 6.8) (bank and post office, in particular). Several use commercial services (shops and grocery stores). Very few use more personal and leisure-oriented services like getting a haircut, working out at a fitness centre, or visiting friends and relatives. The evolution of patterns of work and mobility toward a more articulated use of the services is thus quite evident, although not completely evident for some activities.

It is also quite clear that residents and commuters show a very similar preference to use services during the break. The differences are mainly due to the level of replaceability of the services. For example, some services like the registry offices only serve the population living in specific areas.

Table 6.8 Activities performed during the lunch break (*), percentage of total

Activities	Residents	Commuters	Total
Bank	66.8	53.5	64.5
Mail office	62.0	54.1	60.6
Register offices(**)	41.8	10.9	36.4
Shopping	11.9	15.1	12.6
Foodstuffs shopping	11.1	12.6	11.4
Barbers	8.7	8.4	8.7
Fitness centre	4.4	7.1	5.1
Relations	1.1	3.2	1.5

(*) workers having schedule with lunch-break (**) and others like health services

Source: Dept. of Sociology and Social Research (Univ. of Milan-Bicocca), 1998

Table 6.9 Satisfaction with timetable of the services by time schedules

	With lunch-break	Unsplit	Total
Unsatisfied with public services timetable	43.6	26.2	41.4
Unsatisfied with commercial timetable	16.0	10.9	15.8
Shops should close later	74.0	67.4	73.0
Public office should open during the weekend	88.6	76.1	86.8

Source: Dept. of Sociology and Social Research (Univ. of Milan-Bicocca), 1998

Concerning the level of satisfaction with the actual schedule of commercial and public services, commuters are less satisfied (Table 6.9 and 6.10) than residents. Commuters would prefer longer opening hours, probably because a larger range of time schedules for access to the services fits in better with their pattern of mobility. Also workers whose' schedule includes a lunch break show more dissatisfaction than 'unsplit schedule' workers. The formers suffer for having no time in the afternoon; therefore they ask for a more flexible timing of the services.

Table 6.10 Satisfaction with timetable of the services (*) among residents and commuters

	Residents	Commuters
Unsatisfied with public services timetable	38.0	56.1
Unsatisfied with commercial timetable	10.4	21.6
Shops should close later	65.9	76.1
Public office should open during the weekend	81.8	93.0

(*) workers having 'schedule with lunch-break'

Source: Dept. of Sociology and Social Research (Univ. of Milan-Bicocca), 1998

The main findings of this analysis are the following:

- the structure of labour in the metropolitan area of Milan is still a classical one, with the prevalence of workers having a schedule with lunch break;
- workers use public services often, also during the break, while the more personal activities are mainly performed after work;
- commuters and residents use services during the break at the same rate, with some exceptions due to the replaceability of the services;
- timing and time policy, according to different patterns of mobility, are critical to efforts aimed at improving the quality of urban life.

Urban Governance and Time Policies

Time policies have been recognised as very useful instruments to improve the living standards in a local community. Such policies are based on a preliminary analysis of the mobility and time use of the population and on the time schedule regimes in the public and private sector, viewed as a system of *chronotopos*.

Some administrative experience of implementing time policy in Italian cities has been built up over the last decade (Bonfiglioli and Mareggi, 1997). Urban time policies started in Italy at the beginning of the 1990s within the framework of the public administration reform. New legislation gave the mayor the power to co-ordinate public service timetables, with the aim of revising the schedules to meet the users' needs. This action, in turn led to a change in time schedules regulating human relationships at the level of the city. Since then, several municipalities in Italy have been implementing time policies, sometimes within the overall framework of regional regulation.

Originally public intervention in opening hours meant revising timetables in the field of public services or commercial activities. Today it is applied to a more general framework aimed at a qualitative transformation of public services to achieve higher urban quality. The main issues related to the growing relevance of time policies may be summarised as follows:

- a new organisation of the daily working hours and of the yearly time schedules;
- an increasing mobility of different population groups;
- a greater effort to find new ways of combining work, use of services and care.

Italian experiences concerning urban time policies have been developing according different typologies of intervention:

- the model of users, emphasising differences in gender or life cycle, for a friendlier and more human city. The intervention, which implies the partial revision of timetables in the city, has been adopted in many small towns, including Modena and Reggio Emilia;

- the model of social and collective services, which focuses on the quality of social services for a more efficient city. This intervention has been adopted in Novara and Venice;
- the model of the time plan, a strategic project for a more harmonic city where working time and care time can be both developed. The main experiments were performed in the metropolitan cities of Milan, Rome and Turin.

Four kinds of innovation have been introduced by urban time policies:

- integrated policy design of the calendars of public spaces. This is found in public areas such as urban squares, malls and shopping centres;
- cabling of urban space. This will have a strong impact on the accessibility of public services. It will introduce new dimensions in the interaction between public administration and the citizens. Moreover, it will create new channels and ways of exchanging administrative information. 'Citizens not moving around', tele-shopping, social control and knowledge of administrative action may possibly be the outcome;
- time banks are associations of citizens exchanging fragments of time for small services. This is a renewal of the informal relationships so characteristic of the neighbourhood tradition. The experience with time banks is rapidly growing. They are created directly by citizens with few intermediaries. Time banks offer a solution to problems that are difficult to solve from the institutional side, such as emergencies, quick intervention, work during atypical hours, communication and customer-friendliness;
- mobility agreements are made between social and economic actors. These may stipulate ways to desynchronise starting and ending hours of work and study activities, to improve traffic congestion, or to promote public transportation for certain groups of citizens, e.g. the youth and elderly people.

Developing a Time Plan in Milan

The time plan developed in Milan can be considered a pioneering experience (Bonfiglioli and Mareggi, 1997). In 1990, the promoter of the time plan decided to establish a new office of the municipality devoted to time-related themes. The intervention strategies were based on three levels of action: a political, a methodological and an institutional one. On the first level, the city government designed the general framework, starting from the idea of an overall plan. The main issues on the methodological level were the definition of the fields of action, the choice of topics of analysis and the idea of starting with pilot projects. On the institutional level, an *'Ufficio Tempi'* was set up inside the municipal organisation. This new office was given the task of implementing the Time Regulation Plan (P.R.O., *'Piano Regulator deli Roar'*). The office was run by a steering group at the political level and an operative group consisting of experts and project designers.

Over a three-year period, the Office produced a general methodological draft of the procedures for the definition and implementation of the time policies. The draft plan covered the following items concerning the time schedules and cycles in the city:

- the temporal character of urban life in Milan;
- the legal framework and rules at a national, regional and local level relevant to the time schedules;
- monitoring of the calendar of public services;
- asset allocation of the time of groups of citizens, such as women, children, and elderly people;
- time schedules and time flows of mobility of the inhabitants and of the city users, daily, weekly and seasonally;
- work and study time schedules; shopping and leisure time schedules;
- experimental case studies concerning some issues, like summertime services in the city, or local services in a neighbourhood;
- analysis of waiting times for access to specific services, such as health care.

At the end of this preliminary activity, the office envisaged procedural rules for implementing the time policies. The rules were based on the co-operative discussion among different stakeholders. In the end, they formed the basis for a statement of intent (*'Protocol d'intesa'*). The office suggested the adoption of technical and legal tools for implementing time policies.

The contribution of the Milan office to the foundation of a knowledge base and of procedures oriented to the time policies led to the wide diffusion of time offices in the following years all over municipal government in Italy. They were especially prominent in mid-size cities. These offices promoted a process of legal deregulation of the time calendars in favour of local time policies.

The impact of the office on the actual implementation of time policies in Milan was comparatively low. That is mainly because the unit of analysis and action did not cover the metropolitan area in terms of sociological structure and functional organisation. But their poor performance also reflects the lack of an authority at the metropolitan level, to co-ordinate analysis and policy. Furthermore, the weak implementation is also due to an inadequate analysis of the different life styles and mobility profiles of the people who live in or use the city.

Notes

[1] Istat, Indagine multiscopo sulle famiglie, 1996-1998; national survey on different aspects of the everyday life of the families based on a representative sample of 61,241 individual cases at the national level. The subsample in the Metropolitan area of Milan counts 1523 cases.

[2] Data on daily time schedules and the use of the services are drawn from a survey among a representative sample of 1,300 people living in the Province of Milan. The data were collected by means of mail interviewing by a specialised Poll Institute of Milan (Abacus-Sofres) in 1998. The questionnaire dealt with more topics than the pattern of mobility, like satisfaction with the services, quality of life and perception of time, and the use of technological devices.

Department of Sociology and Social Research (University of Milano-Bicocca), Aspettative e bisogni di chi abita e governa l'area provinciale milanese, supported by the Province of Milan, scientific board: G. Martinotti, M. Boffi, F. Zajczyk, Research report, 1998.

References

Bastian, J. (1994), *A Matter of Time: From Work Sharing to Temporal Flexibility in Belgium, France and Britain*, Avebury, Aldershot.

Bonfiglioli, S. and Mareggi, M. (Ed.) (1997), Il tempo e la città tra natura e storia. Atlante dei progetti sui tempi della città, *Urbanistica Quaderni*, anno 3 (n.12, numero monografico).

Bosch, G. (1986), *Reducing Working Time and Improving Schedule Flexibility*, Report on the Seventh World Congress Iira, Hamburg.

Cafiero, S. and Busca, A. (1970), *Lo sviluppo metropolitano in Italia*, SVIMEZ, Roma.

Camagni, R., Cappelin, R. and Garofali, G. (Eds.) (1984), *Cambiamento tecnologico e diffusione territoriale: scenari regionali di risposta alla crisi*, Franco Angeli, Milano.

Centro Studi San Salvador Telecom Italia (1996), *L'uso del tempo degli Italiani*, Congress Venezia 15 November.

Dhondt, S. (1998), *Time Constraints and Autonomy at Work in the European Union*, EUR-OP, Luxembourg.

Gauvin, A. and Jacot, H. (1999), *Temps de travail et temps sociaux*, Éditions Liaisons, Rueil-Malmaison.

Gershuny, J. (1993), *L'innovazione sociale. Tempo, produzione, consumi*, Rubbettino Editore, Messina.

Gershuny, J. (1995), Uso del tempo, qualità della vita e benefici di processo, *Polis*, IX, 3, pp. 361-377.

Hamermesh, D. S. (1996), *Workdays, Workhours, Work Schedules: Evidence for the United States and Germany*, The W.E. Upjohn Institute, Kalamazoo (MI).

IreR (1991), *Social Survey in Lombardia*, Franco Angeli, Milano.

IreR (1995), *Vivere in Lombardia*, Milano, IreR, Milano.

Juster, F. Th. and Stafford, F. P. (Eds.) (1985), *Time, Goods and Well-being*, Institute for Social Research, Ann Arbor, The University of Michigan, Michigan.

Lavergne, F. de (1985), *Il telelavoro, in IreR-Progetto Milano, tecnologie e sviluppo urbano*, Franco Angeli, Milano.

Manacorda, P. (1996), Tempo e servizi, *L'uso del tempo degli Italiani*, Centro Studi San Salvator Telecom Italia, Congress Venezia 15 November.

Martinotti, G. (1993), *Metropoli*, Il Mulino, Bologna.

Mückenberg, U. (1998), *Zeiten der Stadt*, Edition Temmen, Bremen.

Nussbaum, M. and Sen, A. (Eds.) (1993), *The Quality of Life*, Clarendon Press, Oxford.

Paolucci, G. (1993), *Tempi postmoderni*, Franco Angeli, Milano.

Szalai, A. (1972), *The Use of Time*, Mouton and Co, The Hague.

Treu, T. (1986), Nuove tendenze e problemi del tempo di lavoro. Il dibattito, in Francia, *Stato e mercato*, n. 18, pp. 377-402.

7   Transport and Urban Development: the Potential Impact of Milan and Turin's Crossrail System on the Land-use Structure

PAOLO RIGANTI

Introduction

The 'Crossrails' of Milan and Turin are the most important public infrastructure works built this century in Italian cities. The prospect of new infrastructure investments has led to an attempt to reorganise not only local and territorial mobility but above all to redefine land use and zoning and to transform the main central areas within the metropolitan core.

In Milan and Turin, the aim of the new Crossrail systems is, although in different ways, to promote changes in their urban development. The new Crossrail systems:

I   build an integrated public transport system;
II   improve travelling time and frequency as well as service reliability;
III upgrade and increase long-distance passenger services allowing for swift changes to inter-city main lines running every hour;
IV   connect the railway system to the city transport lines above ground and to the future underground lines.

In spite of the fact that the Crossrails were conceived as solutions to similar problems, such as linking railway networks and improving transport across the city, the scale of the projects differs. Milan was thought of as a metropolitan and provincial project; Turin was envisaged as a project within a national context. Both projects were conceived and developed by the Ferrovie dello Stato, FS - the State railway.

As is underlined in the introduction to this book, the success of any investment strategy depends on how people will use urban infrastructures and on the way in which decisions on investments are conceived and put into action. In this and the next contribution to this book, the Crossrails of Turin and Milan are studied from both perspectives. In this chapter, the Crossrails are analysed from a functional point of view. It notes the impact of the Crossrails on transport demand and assesses its long-term effects, in particular the capacity to give direction to territorial transformation.

In the next chapter, written by Lami, the two projects are analysed from a financial point of view. That chapter shows how the idea of paying for the infrastructure through an increase in land values and the rent generated by the new investments has failed. Real estate reasons have influenced some significant aspects of the project, such as the location of the stations.

It is generally believed that a new transport system has to compete with the already existing system and has to be compatible with the existing territorial organisation. If so, then in evaluating the impact of the new infrastructure on the territory, it is necessary to consider in what way the existent transport system has enabled accessibility within the region and how far this has been developed. In this way, it is possible to identify the areas the new system might affect, compared to others which might be affected less. The conclusion that may be drawn from these observations is that because the new system's infrastructure will have an impact on the organisation of the region, at least two situations should be put to the test. One is an increase in the levels of accessibility; the other is the possibility of further urban expansion. In many urban and suburban areas, these situations have not been tried out, with some exceptions.

As will be seen later, Lombardy still has a certain amount of buildable land. Together with the high levels of car traffic congestion, this supply of land has decreased the accessibility in these zones. Under these circumstances, a new transport system which has been well planned can still have positive effects on the organisation of the region. A good system of mass transit could influence the location choices for residential and industrial sites and thereby induce people to change from private to public transport. Even in the case of the Turin Crossrail, where its influence would be limited to the metropolitan area, it is possible to reorganise urban development in the available areas suitable for building around the intersections of the new system.

In the next section we will present a short review of the literature concerning the relation between transport infrastructure and land use. Section 'Description of the two Crossrail Projects' gives a description of Milan's and Turin's Crossrail projects. The potential effects of Milan's Crossrail on land use and travel are the subjects of section 'The potential Effects of the Milan Crossrail Project on Land use and Travel Demand'.

The Relationship between Infrastructure and Land-use

The presence of any given transit system will impact an area and organise it accordingly. Most of Europe enjoys widespread accessibility, which means any new transport infrastructure has to compete with the existing transport system(s) and take into account how such systems may have already shaped the territorial organisation.

If the above holds true, then when assessing the impact a new transport system has on an area we will have to consider what other existing systems it has to compete with and whether the way they have shaped the area is compatible with the new transport infrastructure.

Each transport system affects accessibility and consequently land use, in different ways. For instance, a new private/individual transport system is going to differ substantially from a new railway line.

Private transport offers more flexibility and speed than other means. In the past, the spread of cars led to a lower density and made suburban areas more accessible. As a result, residential developments, as well as many commercial and manufacturing activities, were relocated (Stover and Koepke, 1988). In an area characterised by diffuse urban development and reliance on private transport, a new proposal for public transport can even have negative effects on activity location. Many urban and suburban areas, which have developed thanks to accessibility provided by the automobile, have insufficient density for a public transport network. In other words, if it appears difficult to replace a private means of transport by a public means, and therefore difficult to stop the phenomenon of scattering solely by a new proposal of transport, it is easier for the opposite to happen. In that case, a polycentric settlement which has always relied on public transport would develop in a dispersed manner, now relying on a new road network.

From the 1970s, with the introduction of public rail transport systems, there have been attempts to organise the land surrounding certain junctions. These interventions would also limit the phenomenon of residential and production dispersal, whether for the efficiency of the transport system or for environmental reasons. An increase in accessibility to certain regional points could have the effect of directing the location of residential and production sites to such places. In this way, a new transport means could become an efficient instrument for controlling territorial transformation. Generally, the aim of new public transport systems is to reduce car travel and improve environmental quality. "The very reason for federal support of rail projects during the past several decades has been transit promise for inducing compact development and the associated benefits of energy consumption and air quality improvements" (Meyer and Gomez-Ibanez, 1981). As we read in an OECD document: "Car travel and the distribution of goods by road have already caused severe problems in cities all over the world and both car use and road haulage are increasing almost everywhere (...) Dispersion encourages more travel by car and necessitates the use of more urban land for roads" (OECD, 1995).

However, in cases of major urban expansion and low levels of accessibility, a new transport system has a considerable impact on the organisation of a region. Therefore, with the careful planning of a new system, it should be possible to control the development of a region. Today this is something which appears to be more and more improbable. Compared to cases such as Los Angeles (Wachs, 1984) and London (Hall, 1988), where the construction of new railway lines at the beginning of the century had decisive effects on the development of the two cities, today a new infrastructure will not result in changes in the levels of accessibility as much as it will in the make-up of the land uses. "In early times, the ability of urban transit investments to strongly influence urban form was indisputable (...) In more recent times, new rail transit lines have been introduced primarily in large metropolitan areas that are already substantially built and that have extensive freeway networks – in other words, in markets with very high levels of accessibility. Accordingly, the impacts of new transit investments on urban form and local markets could be expected to be far less pronounced today than a century ago" (Cervero and Landis, 1993). A less decisive effect of the infrastructure proposal on location choices is also reflected in the lighter burden the transport costs have on the costs of many production activities. As Hall (1995) observed: "The changes in accessibility are likely to be small and not of a sufficient scale to influence location. Transport costs as a proportion of total production costs in many industries are relatively small, and other factors such as skilled labour, suitable sites, availability of government grants and a quality environment are all more important than transport". As "in Western economies, where there is already a dense network of routes, any additional link in a network is only likely to improve accessibility marginally. Other factors such as labour supply, access to markets, availability of land, government grants and incentives are all more important factors in the location decision" (Banister and Lichfield, 1995).

If the above holds true, the 'land-use/transport feedback cycle' as proposed by Hansen in 1959 is no longer applicable in most present-day cases. According to this scheme, a new means of transport, which increases the levels of accessibility to congestion points, could change the make-up of land use (Hansen, 1959). This is no longer valid today, since the levels of widespread accessibility are now higher and the land is to a large extent already built up. "The land-use/transport feedback cycle remains in effect only where accessibility is a scarce commodity (...) this explains why public transport investment alone does not lead to concentrations of development near transit stations" (Wegener, 1995). Findings from research conducted by Cervero and Landis (1993) can also be added to these conclusions in reference to the impact of new rapid-transit underground systems on land use in Washington D.C. and Atlanta: "Around the turn of the 20th century, transit investments generally coincided with a period when such cities were experiencing rapid growth; thus, transit had a strong effect on where that growth occurred. In more recent times, new rail transit lines have been introduced primarily in large metropolitan areas that are already substantially built up and that have extensive freeway networks". For this reason, significant effects are only noticeable in developing countries, where the level of accessibility is low and the possibilities of urban development high. "The empirical evidence now suggests that the accessibility changes are relatively small, particularly in a dense urban network of routes (...) It is only in the developing countries that the changes in accessibility resulting from investment in new infrastructure will have a major impact on regional and local development" (Hall, 1995). If it is true that when costs are the same, the fastest means of transport are chosen, then a new transit system could have significant effects on the means of transport used and hence on the choices of location, only if they guarantee a convincing decrease in the time taken for the movement from one place to another (Holden, 1989). In other cases, change in the levels of accessibility is limited and concerns almost exclusively the main nodes of the new transport systems (railway stations or airports), without being able to reorganise the regional structure (Wheeler, 1981).

Description of the two Crossrail projects

A large share of the transport investments in Europe in the last 30 years went into public transport to link different existing railway lines,[1] the construction of new underground lines linking the railway stations, or the construction of Crossrail systems linking the railway lines directly, as in Munich, Paris, Milan and Turin.

There are two kinds of Crossrail systems: the first one (for instance in Paris) is a new railway line by-passing the core of the city; the second one (like Milan and Turin) is a tunnel allowing different regional trains to by-pass the city. In the latter case, the different trains going through the tunnel make the Crossrail a new underground line with very frequent journeys (Banister, 1994; Banister and Hall, 1981).

The Rail Systems

Milan The Milan Crossrail (MCR) is a new underground rail line connecting the existing rail terminus and allowing regional and national trains to pass through Milan in an eight-kilometre-long and 20-metre-deep tunnel, with two lines running across the city along a NW-SE axis (see Figure 7.1). The cost of the new infrastructure is about 830 million Euro.

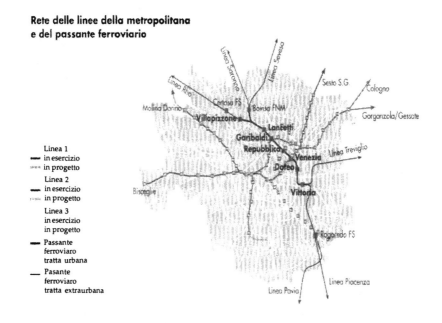

Figure 7.1 Milan Crossrail in the railway network

The rail lines that will use the Crossrail link are mentioned in Table 7.1 (see also Figure 7.1).

The MCR links local stations (the Garibaldi and Vittoria Bovisa Stations). It will connect the regional rail network to Milan,

Table 7.1   Characteristics of rail lines using Milan's Crossrail

Line	KM	Frequencies (Minutes)	Trains/day
Saronno-Lodi	55	30/15	102
Seveso-Pavia	58	30/15	102
Gallarate Treviglio	74	30/15	98
Novara-Piacenza	118	60	36
Novara-Brescia	132	60	36
Malpensa-Bergamo	104	30	68

rationalising the use of facilities and services and linking regional networks in Milan. The aim was to link all the poles of Lombardy to the region's capital city in under one hour's travelling time, with trains running more frequently and cutting travelling time.

The MCR was first discussed in the 1960s by the local governments. It was mainly seen as a city and district link, a sort of metropolitan railway branching outwards, serving the towns and cities surrounding Milan within a radius of 15-20 kilometres. Subsequently, the Crossrail was included in the Regional Transport System and in the Regional Transport Plan. Before the implementation of the Regional Transport Plan, the rail infrastructure system of Milan was split in two different systems: the state railway system (FS), and the north Milan railway system (FNME). The latter has more frequent stations (2500 M), more frequent trains (18 trains per hour) but a lower speed than the FS trains. The FS system is faster, but it is not as diffused as the FNME system. In the past, these two systems were not co-ordinated. The aim of the new Regional Transport Plan is to consider all the railway system in Lombardy as a whole (see Figure 7.2).

Turin   The Turin Crossrail (TCR) includes a new line between two intermediate stations in the core of the existing network. The link is not limited to Turin: it links the entire greater metropolitan area from northbound to southbound (Figure 7.3). The original concept of Turin is to integrate different railway lines (integrating and interconnecting national to regional level), to implement a true regional railway system, and to facilitate the junction with the future high-speed line.

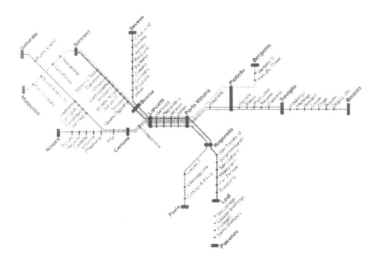

Figure 7.2 The Crossrail system in the 'Servizio Ferroviario Regionale' (SFR) system

The linking up of different railways in the TCR will create the regional network. The network will be managed separately from the rest of the lines, and the number of trains running will make up a de facto city or metropolitan railway. The aims of the new system are the following:

I   to build an integrated public transport system in the Turin metropolitan area;
II to improve travelling time and frequency as well as service reliability;
III to upgrade and increase long-distance passenger services, allowing swift changes to inter-city main lines running every hour;
IV to upgrade freight transport;
V   to connect the railway system to the over-ground city transport lines and to the future above-ground lines.

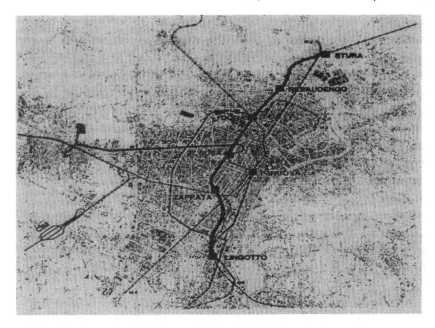

Figure 7.3 The Turin Crossrail in the Railway network

The aims and the mere scale of the project are different in Turin compared to Milan. Turin's Crossrail link is a mere attempt to reorganise the transport node of the city. The city's node consists of its stations, lines and all the railway facilities in the metropolitan area of Turin. There are 18 stations, three engine sheds, and three rolling stock repair centres. Daily passenger traffic amounts to 186,000 trips 75% of which are regional.

Turin has to manage a large number of lines. This has been a problem since the 1950s, as the cohabitation of extremely diversified traffic has led to a crisis in traffic management.

Up till the beginning of the 1980s, there was a progressive demographic concentration compared to the regional area, which had reached 41% by 1981. Between 1951 and the end of the 1950s, the Turin metropolitan area saw increases from 28.1% to 35.3%, for the population percentage in comparison to the region; from 27.8% to 34.2%, for the total activities; and from 41.4% to 46.4% for the authorised industries in the region. The population growth relates above all to Turin, and less to the municipalities within the Turin belt. The tendency to centralise industrial activity in Turin continued in the following years. In 1961 the percentage of workers in the industries in the Turin metropolitan area was 47% of the regional total; by 1971, it had increased to 52%. The 1970s saw the beginning of the critical period for the Turin area, connected with the restructuring of FIAT. The period from 1971 to 1981 shows a demographic decline in the Turin area, while the levels increased in the belt around Turin. The process of industrial restructuring in these years meant a fall in employment for large-scale industry, in part compensated by the increase in employment in small and medium-sized enterprises. This process of industrial restructuring means that previously occupied areas could then form a semi-circle adjacent to the town centre. Along with this process of industrial restructuring and relocation there was a greater dispersal of people throughout the region, despite the evident appeal of the city of Turin.

Since the 1980s Turin has experienced a drop in population and the number of jobs, balanced by an equivalent growth in the surrounding municipalities. Major industrial reorganisation resulted in the growth of small and medium-sized enterprises. The outcome has been that services to both residents and industry relocated outside the metropolitan area, while the service industry itself on the whole remained in the central area. Mobility has therefore increased and is more common in the metropolitan area.

Turin differs from the Milan project in that the aim in Turin is not to maintain a multi-centred system in the Piedmont region. Rather, the aim is to improve access to Turin and enhance its central role compared to the smaller centres in the region.

The Political Tools

With the Crossrail project, Milan experimented with a new planning tool, called Documento Direttore del Progetto Passante (DDPP). Crossrail was conceived as a master project, developing features of the master plan without redefining its objectives. The MCR was also seen as an opportunity to organise a number of major urban projects and thus as a means to guarantee the effectiveness of a number of municipal projects.

With the DDPP, Milan has tried to manage the changes taking place in the areas affected by the new facilities. The DDPP introduces the notion of area project ('progetto d'area'). An area project is defined in the DDPP as the implementation of the political and administrative indications of the DDPP. It is supposed to inform the architectural design for the various projects, promoting agreements between private and public actors. The most important projects include some huge urban transformations (such as the Technological District at the Bicocca Pirelli and a service area near Linate Airport (Montecity).

Other projects include:

I   re-use of the former industrial areas;
II real estate development;
III service and facility networks.

In 1995 eight areas were identified for feasibility studies in Carnate, Como, Monza, Pioltello, Saronno, Varedo and Varese. Projects and protocols have in the meantime been signed as drafts for the agreements ('accordi di programma').

The Turin municipal authorities view the building of a new railway line crossing the city as an opportunity to upgrade the urban environment along the railway itself. This is why new lines are being built underground wherever possible and existing ones are being covered. This will make it possible to connect previously separated sections of the city and to enhance surrounding areas in redesigning the city. The new master plan (Piano Regolatore Generale) views the Crossrail as the backbone of change in the city centre. In the late 1980s, the new master plan moved implementation of the Rail Plan forward so that it could serve as a trigger for a major urban renewal. The renewal was envisioned for areas and zones emptied by industry: the Dora steel works, the Valdocco and Porta Susa railway junctions, the former prisons, the railway workshops and the Materferro areas near Largo Orbassano. Lastly, there are large areas mostly zoned for services: the development of the polytechnic (Faculties of Architecture and Engineering); the creation of the Dora Technological District (Polo Tecnologico); and the creation of two city parks, one near Dora and one near Porta Susa. The Crossrail link will have a major impact on the central areas of the city. The new supply of public transport will greatly increase levels of accessibility in an area virtually saturated with private traffic. One can therefore reasonably suggest that the new infrastructure will contribute to the urban reorganisation of Turin, albeit only in these areas and not according to the Milan pattern.

The Potential Effects of the Milan Crossrail Project on Land-use and Travel Demand

Over the past few years, Milan, like Lombardy as a whole, has been replacing public transport with private means. This has resulted in two new problems:

- urban sprawl and land use have increased scattering residential and production developments outside the central areas;
- private transport is saturated and traffic has peaked, so that the existing road network is now unable to cope with demand.

If these problems could only be addressed by extending the road network capacity, it would result in a larger area of land being used. This would only partially solve the existing mobility problems, and even then only for a period of time. Rail transport appears more promising, both as far as system efficiency is concerned and to safeguard Lombardy's multi-centred pattern.

Location decisions are not as dependent on travel accessibility levels as they were in the past. Between 1951 and 1971, in Lombardy, the rail-served population increased by about 37%, and between 1971 and 1991 the population fell by about 4.5%.

Milan and Lombardy have witnessed an increase in out-of-town transport, the flow being mainly carried by private means and by a lower degree of self-containment.

In the years 1981-1991, demand for out-of-town transport increased by at a rate of 3% per year in Lombardy, while the share of public transport declined from 38.9% to 30.4%. Rail transport remained stable at 14.9% (Table 7.2). The increased demand for out-of-town transport corresponds to a decline in the self-containment rate in the various municipalities. Mobility has, therefore, increased and larger towns appear to act as magnets for distant municipalities, thereby enlarging the commuter belt. It is not just, or not so much, a matter of people – their number is relatively stable, with a 1.2% yearly growth from 1981 to 1991 – but more a question of the type of transport used and the distances covered. Larger district capitals (Capoluoghi) have ceased to grow, while their surrounding areas are growing and more poles are developing. Increased out-of-city and private transport are mirrored in residential patterns across the region as a study of population shifts shows. That study relates the demographic shift to the presence of the railway from 1951 to 1971 and then from 1971 to 1991. In the earlier period, population increased nearly twice as quickly in municipalities with a railway. The trend reversed in the following 20 years (-4.3% in municipalities with a railway) compared to an 18.3% growth in those without a railway.

Table 7.2   **Changes in modal split in commuting travel in Lombardy** (1981-1991) (%)

	Public transport	Rail transport
Milano	-8.7	+0.9
Bergamo	-8.6	-0.9
Brescia	-8.3	-0.1
Como	-6.9	-0.2
Cremona	-9.6	-0.2
Mantova	-9.4	-0.7
Pavia	-10.1	-1.3
Sondrio	-10.0	-1.7
Varese	-6.2	+0.4

Trend reversals in residential development patterns may be partly attributed to the relocation of a number of production and manufacturing activities. When these were moved out to the outer towns, the resulting commuting was not always backed up by adequate rail infrastructure. This is another reason why the transit system expanded and spread over the region using its road network. A large portion of Lombardy can still be built up (25%), and this requires a compromise.

In the long run, closely inter woven public and private systems are needed to interrupt a process which has led to excessive traffic and environmental damage. In 1982 the Regional Transport Plan suggested the creation of a Regional Rail Service to dove-tail with the national rail service and the Milan rail service at a time when the only way to address a growing demand for mobility appeared to be 'more roads'. The aim was to shift part of the private motorists to the public system. The MCR would improve mutual accessibility between and among cities, the region and the metropolitan area (Table 7.3). New stations in the underground section would be created so as to improve service, sharing passenger loads throughout the urban area and maximising the use of the existing rail networks. Road transport could then be used to reach the closer stations: large areas of Milan could thus be 're-thought', and not just re-conceptualised in terms of the new transport system. If local trains ('diretti') are compared to the future regional trains, the following improvements can be expected: more frequent service (95% more frequent), and faster travelling time (14% faster).

Table 7.3   Changes in accessibility from Milan: variations 1992-2000 (%)

	Travel time	Train frequency
Bergamo	-18.5	126.3
Cremona	-15.8	48.6
Pavia	-12.1	59.0
Novara	-12.1	36.5
Como	-9.1	24.6
Varese	-6.3	13.3
Brescia	-4.6	6.8

The MCR is not just a means to make the Milan and Lombardy transport systems more efficient; it embodies hopes that the existing lay-out and land use will hold. The hope is that Lombardy's multi-centred structure will favour the location of production and manufacturing activities close to the regional rail nodes, thus avoiding traffic problems. In fact, those were the problems that in the past led to a development pattern based almost entirely on private transport.

Conclusion

On which conditions can the two projects be effective in achieving their targets? Both Milan and Turin have been considering new Crossrail systems since the 1970s, to integrate and join the railway lines converging on their city centres, thus making through transit possible. Since the 1980s, these projects have also been seen as an opportunity to regenerate and upgrade some of the city districts using the new accessibility as a springboard for development. The impact that the two new Crossrail systems could have on the organisation of the regions of Lombardy and Piedmont have been examined. In particular, it has been underlined that the rapport between a new infrastructure proposal and the reorganisation of a region should not be taken for granted. The congruence will depend on numerous factors; most importantly, it will depend on the degree of accessibility present in an area and the availability of buildable land.

Milan is working mostly on vacated areas, while Turin is focussing on areas to be freed by the new Crossrail system. In both cases, these projects are considered to be of strategic importance, involving several actors. They appear to be comparable, in spite of their differences. The cities also see the Crossrail project as an opportunity to manage change in a more general framework, rather than as single episodes in an urban environment. The two cases illustrate the strong impact these infrastructural and facility-based actions can have on the overall city. The transport system is therefore seen as a lever for change, whereby the authorities are able to make major transformations while governing them in a general framework.

Notes

[1]   This is also a major aim of the European Commission transport policy (European Commission, 1996).

References

Banister, D. (1994), *Transport Planning, in the UK, USA and Europe*, E & FN SPON, London.

Banister, D. (Ed.) (1995), *Transport and Urban Development*, E & FN SPON, London.

Banister D. and Hall, P. (1981), *Transport and Public Policy Planning*, Mansell, London.

Banister, D. and Lichfield, N. (1995), The Key Issues in Transport and Urban Development, in D. Banister (Ed.), *Transport and Urban Development*, E & FN Spon, London. pp. 1-16.

Cervero, R. and Landis, J. (1993), Assessing the Impacts of Urban Rail Transit on Local Real Estate Markets Using Quasi-experimental Comparison, *Transportation Research*, 27A, pp. 13-22.

Hall, P. (1988), *Cities of Tomorrow*, Blackwell, Oxford.

Hall, P. (1995), A European Perspective on the Spatial Links between Land-use, Development and Transport, in D. Banister (Ed.), *Transport and Urban Development*, E & FN Spon, London, pp. 65-88.

Hansen, W.G. (1959), How Accessibility Shapes Land-use, *Journal of the American Planning Association*, 25, pp. 73-76.

Holden, D.J. (1989), Wardrop's Third Principle – Urban Traffic Congestion and Traffic Policy, *Journal of Transp. Economics and Policy*, 23, pp. 239-262.

Meyer, R. and Gomez-Ibanez, A. (1981), *Auto, Transit and Cities*, Harvard University Press, Cambridge.

OECD (1995), *Urban Travel and Sustainable Development*, OECD Publication Service, Paris.

Stover, V.G. and Koepke, F.J. (1988), *Transportation and Land Development*, Prentice Hall, Englewood Cliffs.

Wachs, M. (1984), Auto, Transit, and the Sprawl of Los Angeles: the 1920s, *APA Journal*, 50, pp. 297-310.

Wachs, M. (1993), Learning from Los Angeles: Transport, Urban Form, and Air Quality, *Transportation*, 20, pp. 329-354.

Wegener, M. (1995), Accessibility and Development Impacts, in D. Banister (Ed.), *Transport and Urban Development*, E & FN SPON, London, pp. 157-161.

Wheeler, C. (1981), The Land-use Impact of Rapid Transit in Toronto, *Essay for the Degree of Bachelor of Environmental Studies in Urban and Regional Planning*, University of Waterloo, Ontario (CAN).

8 Transport Infrastructure and Planning Policies: the Importance of Financial Analysis in the Crossrail Projects Milan and Turin

ISABELLA LAMI

Introduction

In view of the way plans are implemented in a number of cities, it is obvious why major difficulties arise when the plan has to be put into action. In Italy, the *'piano regolatore'*[1] (PRG) is conceived prior to development; the actors, modes and timing of its implementation are unknown.

Zoning is often wrongly believed to be a model for the market, as if it really had the power to change the image of its urban market. The problem is that very often there is no correspondence between the forecasts in the plan and real demand in the property market.

The case study that we present here is a typical one in the sense that there is no correspondence between the plan and reality. What makes this particular case so interesting is that it is a case where urban planning decisions interact with a major infrastructure project.

The point of departure for this study is the observation of what probably constitutes the greatest infrastructural work of the century for two Italian cities, the Crossrail systems of Milan and Turin.

The new urban development plan started up a great urban

renewal, and the first project is the Crossrail in Turin. The project covers almost three thousand square metres. The area slated for renewal is largely made up of old industrial sites, depots and railway offices. These properties will be valued and put on the market.

The Crossrail in Milan is now almost finished. That plan envisions the creation of new stations, around which major public and private property projects are being developed.[2]

These two projects have been in place for over fifteen years. Nonetheless, in both cities, the projects have run into difficulty in taking off due to financial reasons. Besides public funding, the vast availability of space connected with the realisation of the Crossrail seems to be out of proportion to market conditions. In fact, the market's reaction is now revealing itself to be quite different from the initial expectations.

To analyse the reasons for this situation, we have divided this chapter into two parts.

In the first part 'The Crossrail projects', we analyse the financial plausibility of the Crossrail in Milan and Turin. For both cities we will look at specific issues – the objectives and the promoters, the costs and how they are covered, and the value of the areas – in an attempt to understand the financial burden of the choices in an urban transformation project.

In the second part 'Financial Analysis of Turin's Crossrail', we focus on the city of Turin. We have applied financial analysis twice to one of the zones of transformation of the city. The first time we used the information extracted from the PRG. The second time, we used information concerning the revision of the city plan, which is currently under discussion.

The first application of Discounted Cash-Flow Analysis to this zone highlighted certain critical elements of the PRG (such as the dimensions and uncertainties of the costs of rehabilitation, the effect of ceding areas to the public authority, the excessive tertiary use). The way events subsequently developed in Turin proved the accuracy of these results.

The second analysis demonstrated how the changes that will probably be made to the plan will greatly improve the situation.

In this context, we will highlight the role of property interests. From the beginning of the project, they have not only served as a financial instrument but, more importantly, they formed one of the motives for action by the principal players and promoters involved.

The case study of Turin will show how profitable (if not necessary) it is to supplement planning and regulatory controls with a forecast of the economic and financial aspects of the operation. This is important so as not to neglect elements that are essential to an understanding of the urban situation. And that understanding is indispensable when the plan is being drawn up. In addition to determining the expected profitability of the project, assessments of this type provide important information (even if not completely objective, at least obtained from declared assumptions). That information, in turn, can constitute a basis for discussion among the various players involved, also during negotiations.

The Crossrail Projects

The Crossrail represents a considerable portion of the overall investment of fixed urban and regional capital in a nation like Italy. There, an incremental policy is usually applied, which entails small adjustments and partial modifications.

The Crossrail means a total cost to Turin and Milan of almost 1500 million Euro. For Italy, an infrastructure project of this size is rather unusual. Investments of fixed capital are often limited. The policy is to spread the funds among other matters of public interest and to ensure the even distribution of infrastructure (Secchi, 1996). The problem is that the lack of funds is 'diluting' this operation in the course of time, this creates the risk of weakening its effectiveness.[3]

The 'time' element is crucial. The cost is one reason: "the abnormal length of time needed to complete Milan's Crossrail means an increase in expenditure by 50%" (Pugno, 1994). Another reason is the role that this infrastructure is supposed to play.

Because of the continual postponement of the realisation of considerable parts of the project, the results end up being obsolete. The change in conditions from when the project was created then compromises its effectiveness when faced with new and unexpected challenges.

An intervention of such dimensions has a huge potential for change at a territorial, location and property level. Thus the project requires not only a huge amount of investments, but also close co-operation between the different public bodies. The combination of strong funding and strong co-operation is necessary so as not to restrict it to being 'a railway operation', but enabling it to become an effective strategy for urban and regional transformation.

There is general agreement on the fact that increasing accessibility should be one of the prime goals of transport investment. In many cases, however the presumed capacity of a project to generate other positive impacts is seen as the main motivation for undertaking such investment.

As far as property prices are concerned, studies conducted on cities in North America have shown limited effects.[4] The empirical evidence does not show beyond a doubt that transport investment in all areas leads to increased development activity.[5] Whilst the existence of some link between the two phenomena cannot be denied, the nature of this relationship demands further empirical evaluation (Grieco, 1994).

Objectives and Promoters

The Turin crossrail At the beginning of the 1980s, the Ferrovie dello Stato (FS - State Railway), in agreement with the Region, decided to reorganise and reinforce the Turin node, so as to adapt it to the national rail system and create a fully interconnected network system. The forecast cost was about 510 million Euro, with works beginning upon approval of the ministerial budget.[6]

At the same time, the number of disused industrial areas (which, not by chance, were located along the railway tracks), together with major infrastructure works, represented a unique opportunity for Turin to upgrade its urban environment. In this manner following negotiations between the municipality and the railway company, it was decided to create a new link in the tunnel between Lingotto and Porta Susa stations. This would quadruple the tracks between the latter and Turin's most northern station (Stura). A decision was also made to create two new stations and insert another line, run by a franchise company, into the FS network.

The quadrupling of the tracks would, however, mean further aggravating the caesura, already present in the inner city, as a result of the railway cuttings (physical barriers created by the tracks, Figure 8.1). On the municipality's request, it was therefore decided to extend the process of lowering to all four tracks, with the costs being almost totally paid for by the city. The covering of the tracks would allow the parts of the city that had been separated by these 'trenches', on the north-south axis, to be rejoined.

Figure 8.1 Physical barriers created by railways in Turin

In theory, investments in transport provide accessibility advantages for the sites served, as these benefits could be capitalised in higher revenues. In general, however, the capitalisation of the effects of rail transport seem highly localised and linked to the context. At present, the greatest progress made to recover some of the benefits produced by public transport investment has been through forms of partnership connected to local property projects. "While value capture is a conceptually elegant mechanism for co-financing public transport investments, in practice it is fraught with implementation difficulties" (Cervero and Landis, 1995).

Beginning with the exclusive purpose of providing transportation links, the Crossrail then went on to serve multiple aims. In addition, it became an instrument for the property market. In this way it combined the destiny of railway areas (both those that were disused and those won back by covering the tracks) with the new town-planning scheme for the city. The FS, therefore, appeared as a powerful new group on the Turin scene (Garelli, 1993).

Figure 8.2 The new main road that will cover the line beside Porta Susa Station

One part of the master plan (in Italian, the '*Piano Regolatore Generale*' or PRG) called the 'Spina Centrale' winds along the underground tracks. On top of the tracks, the plan foresees the creation of a new key linear area, with the realisation of a north-south throughroad (Figure 8.2).

It should be emphasised here that an important FS area, a part of the 'Spina', is the centre of one of the great transformations in the city. It is the site of the second part of the Polytechnic, set up in the area of the railway workshops. That site was chosen following heated discussions on the alternative of locating it in the area of a former steel plant situated elsewhere. In addition, there is a series of building plans to provide premises for the tertiary and hotel sector within the Porta Nuova railway depot. These enterprises fall under a precise real estate strategy through the national 'Trenoporto' plan.

The 'Trenoporto' project was launched in 1988. It was intended to increase the value of the railway stations (central places within the city and therefore ideal for commercial, cultural and tertiary projects). At the same time, it was supposed to increase the revenue of FS real estate. The programme concerned 52 stations in 28 cities, amongst which was Turin.

For the areas recovered from the railway cuttings, Turin's town-planning scheme attributed a higher building capacity (0.7 sqm GFS/sqm[7]), a capacity which could also be transferred to other areas. About 16% of the buildable area, as foreseen by the 'Spina Centrale', belongs to the FS. These have traditionally been considered unbuildable areas. The railways were thus returned to being real estate property. Of course, that property was located along an infrastructure which was decisive in the development of Turin. All of a sudden, it became buildable, which increased its value greatly.

It is evident how the power of the FS is increasing in the negotiations with the local administration. The FS thanks its new standing to its double role: it owns numerous strategic areas in the city and is being directly involved in the most important infrastructure operations for Turin (Chicco and Saccomani, 1993).

The Milan crossrail The Crossrail project for the capital of Lombardy was launched, like that of Turin, at the beginning of the 1980s. The aim was to create an integrated, united system across the city.

The agreement for the realisation of the project was signed in 1984 by the City of Milan, the Lombardy Region, the FS and the North Milan Railways (FNM). The agreement was made at a time when the potential of the Milan node was virtually at a saturation point. The objective was to connect the railway network of the south-east region (served by the FS) with the north-west one (managed by the FNM), by means of a direct underground link between different stations. That connection was intended to transform the city's railway system from 'terminus' to 'Crossrail'.

The Crossrail project is a child of the times. It reflects the logic of the years in which the transformation of the city took place. According to that logic it was necessary, above all, to create conditions for good accessibility. That meant shifting attention to the realisation of infrastructure and services of a high technological level; these were considered attractive elements for business (Fareri, 1990).

In Milan, like Turin, the restructuring of the railway system was accompanied by a desire to find a new use for the disused areas along the Crossrail route. However, the upgrading of the city was more gradual. Urban renewal was not the essence of the town-planning scheme as it was in the capital of the Piedmont. It should be noted that, out of the four stations that were created *ex novo* – i.e., not linked to the underground stations or other working railway stations – two are located in the FS area. The Lancetti Station was constructed near railway yards; the new station, Vittoria, whose construction was long discussed, replaced an old abandoned railway station.

There was a clearly defined strategy for promoting FS property towards the metropolitan areas, as we have seen in both Turin and Milan. The Metropolis Company was established for the purpose of managing and promoting railway real estate and property. The reorganisation of the public transport network and the simultaneous redesign of parts of the city means a strong increase in value of the railway areas, located between the historic centre and the suburbs.[8] The strategy of the FS is to put the railway/urban infrastructure relationship into a new light. The FS does not portray the railway structures as an obstacle for the development of the city. Rather, they are seen as distributors of traffic and functional infrastructures specialised in easing congestion along the connecting network, optimising the mobility between the areas of interest.[9]

Such a strategy recalls the action of other European railway associations. One thinks of the transformation of Broadgate, in the city of London, where an agreement between two private developers, British Rail and the local borough, resulted in an entire re-definition of Liverpool Street Railway Station. The tracks were laid underground; a small railway Crossrail system was made to avoid intersections across the city routes and 400,000 sqm of offices were built in the vacated areas. The refurbishment of Liverpool Street Station was financed thanks to the property operation that was launched at the same time. Vice versa, the property operation was able to benefit fully from the better accessibility of the site (this is one of the features that allowed areas to be rented rapidly and at high rates).

The working of Milan's Crossrail – made up of an eight-km-long tunnel with six underground stations and one on ground level –

should allow an integrated regional service coherent with the polycentric development of the region, connected with Milan's underground transport system. This major infrastructure work thus intends to create transit circulation. Above all, it is intended for district and regional travellers. In that sense, it differs from that of Turin, which should influence, above all, the residential possibilities of the metropolitan areas. In Turin, the aim is to make the rail networks more accessible.

The Costs and how They are Covered

The crossrail in Turin The total cost of the Turin Crossrail should prove to be around 670 million Euro.[10] The agreement between the FS and the Municipality of Turin foresees that all works of a strictly transport nature would be carried out by the railway company. The costs related to the coverage of the tracks would be the city's responsibility (these costs were estimated at about 240 million Euro).

While the FS could make use of state funds, there was a problem as far as the municipal resources were concerned. The work to be carried out by the city would, in part, be financed by loans provided by the state. The assumption was that the income derived from the plan's interventions would provide the municipality with the necessary capital for these works.

There is a deep-rooted conviction that 'empty towns' are in reality 'full of revenue' (and therefore not just opportunities to upgrade the city). In that light, the PRG envisaged a high quota of building for the zones to be transformed (0.7 sqm GFS/sqm) and the same for all the areas, hoping to trigger a huge volume of real estate business.

The 'Spina Centrale', however, had a peculiarity: 10% of the 'superficie territoriale'[11] had to be surrendered to the city as service areas to satisfy existing needs. This provision was added to the 'standards'[12] envisaged by the Regional Law. Those standards were already strengthened by the plan's definition of streets and squares as buildable areas belonging to the municipality. However, those were sites on which they could obviously neither build nor meet the 'standards'. In this case, an underground solution was required, naturally at higher costs.

The plan tried to avoid resorting to compulsory purchase orders. It did attempt to reach the amount of services necessary by means of the new transformation. Thus, the real estate companies were burdened with these extra services. That raised the risk of economic problems, if it didn't make the project completely impossible.

The crossrail in Milan   The estimate, updated in October 1996, showed a cost of about 850 million Euro[13] for the urban section of the Crossrail. The initial agreement divided the costs in three for each party (the Region, the Municipality and the FS). The actual agreement saw a subdivision of the costs for the Region and the Municipality at 50% each. That money would be applied to construction, public facilities and railway equipment for the urban section. Works on the external area and the cost of railway fittings would be covered by the state Railway Company.

In the initial project, the Vittoria Station was intended to be at ground level. It was to be part of the outer urban section (and therefore the responsibility of the FS). However, in 1996 it was inserted into the urban section as part of the underground. It thus became the responsibility of the Region and the Municipality.

It turned out to be particularly difficult to maintain the balance between cost and funding. The problem lay in the amount of time that elapsed between the formal decision on the funding and the effective availability of capital.

The 1988 Finance Law foresaw a temporary delay of 653 days. However, the Finance Law of 1990 delayed payment of the first part of the funding by 1,457 days. This meant a noticeable rise in the implementation costs. The increase was linked above all to the interest charges, which rose over the period needed to carry out the work (18-20 years compared to the predicted 7 or 8).

Financial Analysis of Turin's Crossrail

To try to understand the reasons for the deadlock of the Turin PRG, we applied a form of financial analysis, namely *Discounted Cash-Flow Analysis* (see the note on this in the appendix), to one of the four development sectors proposed by the plan, the 'Spina 1'. This area is the final part at the southern end of the works carried out to transform the Spina Centrale (Figure 8.3).

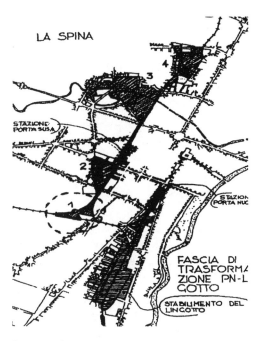

Source: Indovina, 1993

Figure 8.3 Draft of Spina Centrale

Planning and Building Characteristics of Spina 1

The area is currently occupied by a complex of single-story buildings. Until a few years ago, they were used for manufacturing automobile components and stocking cars. If this is the actual situation, then work to recover the site should not be particularly expensive. Meanwhile the covering of the Crossrail tracks has already largely been completed. The planning and building regulations for implementing the project establish the building capacity and the areas to be used for services (table 8.1).

Table 8.1 Planning rules for Spina 1

Maximum territorial index (sqm GFS/sqm ST)		0.7
GFS by type of use :		
Housing	Min.	38%
Business and personal services	Max.	7%
Offices	Max.	55%
Total service areas (ST) (min)	182.000	sqm

Source: Norme urbanistico-edilizie di attuazione del PRG

This table refers to one of the four sectors of the Spina Centrale. It allows us to highlight two significant elements of Turin's PRG:

* the index of building capacity;
* the high supply of property for offices.

This building capacity index of 0.7 sqm GFS/sqm ST was applied to all sectors of the Spina Centrale in order to implement an equalising policy. In reality, this effect has not been achieved. There are three reasons for this.

In the first place, the areas have different characteristics. Not all of them need the same amount of work to re-use them. There is an inequality of expenditure, even though the PRG treats them equally.

Secondly, the transformation of each zone of the Spina is governed by the outcomes of a single study of the area. Through the formation of a 'comparto',[14] the owners involved in the implementation of the PRG share the costs and benefits deriving from the works realised according to their individual ownership shares. Thus, the costs and benefits are calculated independently of the final use assigned by the plan for each plot of land. To make the type of project planned for their land the same for each owner, the new plan for Turin thus uses the *comparto*. This encourages the citizens to respect the indications given in the regulations for the zone. This applies not only to alignment but also to the use of the buildings, the number of floors, and the location of the areas for services. The problem is, however, that it is rarely possible to get agreement between the different owners of the sub-zones. And agreement is needed in order to start the operations of transformation.

The third reason why an equalising policy has not been achieved is that the differences in values in the urban fabric – differences created by the market structure – lead to very different transformation values for the various sectors. The discrepancy is most evident between land in almost central areas and land in the outskirts. The consequence is a considerable diversification of land purchase prices. The wide price range conditions the profitability of the investment with a disproportionate impact.

In this way, the equalisation that the plan intended to attain through the uniformity of its planning parameters ends up creating a general state of inequality. This apparent inequality seems to risk blocking all transformation operations. Or it seems that only the strongest operators manage to benefit from this index of building capacity.

Furthermore, it should also be noted that this coefficient is high. In fact, it is so high that in some sectors this theoretical building capacity would have to be transferred. In other words, it would be necessary to build on different land, not on the plot that had generated the GFS (Gross Floor Space).

As far as the second point is concerned, the total supply of built property generated by the Spina Centrale is almost two million square meters. Of that total, about 40% is devoted to housing and 30% to offices. This proposal concerns a city whose property market is stagnant, and not only for macro-economic reasons. This crisis has, in fact, structural features. In particular, the resident population is declining and the number of owner-occupied dwellings is relatively high.[15]

Given these forecasts in the variation of the population, it is unlikely that the property market will return to previous levels. It is more likely, instead, that the demand for housing will be expressed as a demand for quality.

As far as the offices are concerned, the share that the plan envisages on the Spina Centrale is equal to 80% of the city's current public offices. In this respect the PRG very probably suffers from the great trust that had been placed in the development of services. Such trust was common at the time when it was drawn up. However there has been no boom in this sector. This means that the proposals in the plan find no response in the dynamics of the Turin property market. Thus, the large quantity of property offered by the PRG is running into numerous problems in terms of implementation.

Looking at the spaces of Spina 1 destined to services, one of the problems common to all the sectors of the Spina Centrale is evident (Table 8.2). In effect, the peculiarity of the Spina Centrale is the obligation to cede some of its territory to the city, in the event of transformation of the area. To satisfy the city's requirements, it has to surrender areas for services amounting to 10% of the ST.

Table 8.2 Area for services in Spina 1

	GFS	sqm for services		
		art. 21	city	Total
Quantity (sqm)	135,053	104,188	77,812	182,000

Source: Studio unitario dell'ambito 1- Spina centrale

The plan has tried to avoid any recourse to compulsory purchase. At the same time, however, the plan has tried to provide the quantity of services needed by the city through the new transformation. The property operations are thus burdened with these additional services. They increase the risk that these projects will no longer be economically attractive, if not completely unworkable.

In the case of Spina 1, the situation is even more serious. There, the quantity of services demanded is fixed, amounting to more than 10% of the land area. This obligation comes on top of the standards laid down in the Regional Law.[16] Together, these requirements give rise to a very high percentage of services. The share becomes even greater when building rights are transferred. That is because there is no transfer of the related service quotas, which must then be provided in both zones. In practice, this means that the services must be doubled.

Even the standards in the Regional Law, calculated on the basis of the quantity built, are actually increased 'compulsorily'. In fact, as mentioned above, the Plan defines squares and streets belonging to the city as buildable areas, even though nothing can be built there.

The situation is not improved by the fact that it is not often possible to provide the standards demanded above ground. Thus, most of the car parks, which represent a significant amount of the services to be provided, have to be built underground, at decidedly higher costs.

The legal status of the area is relatively simple. The three owners are: the city, with a land area of 40.13%; Materferro, a FIAT Group company, with 42.16%; and the state Railways (FS), with 17.71%. In reality, however, it is not feasible to put up the buildings on the areas owned. The problem is that is not feasible, as the land belonging to the city is already in use for the existing roads, while and that of the FS is in use for the railway tracks. Part of the site is already covered and part is still in cuttings. However the PRG has attributed the same building rights to these rail and road corridors as for the areas to be transformed. The Materferro plot is thus the only area on which all building work will have to be concentrated.

In addition, the transformation of the area is bound by the plan to the formation of a unitary study of the zone extended to the whole of Spina 1 (Figure 8.4). For this area, an 'urban improvement

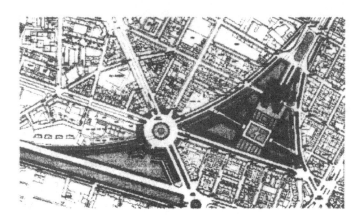

Source: studio Mellano, Torino

Figure 8.4 PriU di Spina 1

Table 8.3   Discounted Cash Flow Analysis to Spina 1 (PRG data)

Costs	%	Price (lire)	Quantity	Unit	Total
Real cost suitable areas		560.000	90.659	mq SLP	50.769.040.000
Technical expenses	5				7.115.425.000
General expenses	4				11.741.172.395
Commercial expenses	2				7.206.328.000
Total					76.831.965.395
URBANISATION COSTS					
Construction					
Housing	5				3.559.575.000
Business and personal services	5				541.240.000
Services	10				6.029.220.000
Primary Urbanisation					
Main sewers		150.000	39.706	m²	5.955.900.000
Network infrastructures		75.000	99.708	m²	7.478.100.000
New roads in the project		80.000	10.234	m²	818.720.000
Artificial standards					
Underground car parks		580.000	35.913	m²	20.829.603.000
Secondary urbanisation					
Gardens		120.000	34.000	m²	4.080.000.000
Foot-path		150.000	12.495	m²	1.874.250.000
Square development		280.000	7.769	m²	2.175.320.000
Fountain		150.000	314	m²	47.100.000
Induced urbanisation					
Railway track covering		2.961.000	3.398	m²	10.061.478.000
Total					63.450.506.000
CONSTRUCTION COSTS					
Residential building		1.500.000	47.461	m²	71.191.500.000

Costs	%	Price (lire)	Quantity	Unit	Total
Business and personal services		1.400.000	7.732	m²	10.824.800.000
Offices		1.700.000	35.466	m²	60.292.200.000
Total					142.308.500.000
TOTAL COSTS					-282.590.971.395
REVENUES					
Residential building		4.000.000	47.461	mq	189.844.000.000
Offices		4.000.000	35.466	mq	141.864.000.000
Business and personal services		3.700.000	7.732	mq	28.608.400.000
TOTAL REVENUES					360.316.400.000
Interest rate		Annual 5.89%	Six-monthly 2.90%		
NPV					-43.349.653.260
IRR ANUAL					-0.38%
IRR SIX-MONTHLY					-0.19%

plan' (*Piano di Riqualificazione Urbana – P.Ri.U.*), has been introduced. This is a tool introduced by law no. 179/92, which aimed at promotes mixed private/public projects through ministry funding and the slimming down of approval and implementation procedures. The P.Ri.U. concerns the building rights of FIAT and the City of Turin. It is on the basis of the figures given in the P.R.i.U. that the Discounted Cash-Flow Analysis was conducted.

The Discounted Cash-Flow Analysis Applied to Spina 1: PRG data

The first application of the Discounted Cash-Flow Analysis to Spina 1 used the data of the PRG (Table 8.3). The duration of the project was considered to be sixteen semesters, from the time of acquiring the land to the conclusion of sales. This is an optimistic hypothesis. In reality, it presumes a more positive trend in the property market than the present one. But it is justified by the renewal of the city through these major urban developments. A research centre, Finpiemonte, conducted a study for the City of Turin (November 1996) that consisting of an economic-financial simulation of the P.Ri.U. of Spina 1. The study indicated the time of the operation (from permission being given to the end of sales) as about four years.

The first cost item concerns the land. This is another critical element of the Turin PRG. In fact, it was decided to designate as buildable areas parcels of land which in reality are not suitable; given that the disused industrial areas demand costly, major improvement work before they can be built on.

The designation of land as industrial zones or 'buildable areas' gives them a theoretically high market value, derived from the construction possibilities conferred by the PRG. But this is not their real value. To determine that, it would be necessary to deduct the costs of demolition and improvement. Those vary according to the previous use of the sites (industries with pollution of varying levels, the presence of asbestos and so on.) Moreover, those costs are not easy to quantify. Often the owners themselves do not even have a thorough understanding of the production methods used in these industries.

On this point, one could object that it will be up to the market, through the mechanism of supply and demand, to set the right price for these plots. But it should be recalled that the ownership is represented by a few large public and private companies; on their books, they already discount the (theoretical) values deriving from this building capacity. In that light, it is easy to understand the importance of the value to this land attributed by the PRG. Faced with these problems, Finpiemonte (1997) uses the term 'suitable area'. This is defined as the property value net of the costs of demolition and rehabilitation. The rationale behind this correction relative to the theoretical market value is to define the real conditions of feasibility. The underlying assumption is that the costs of rehabilitation will be borne by the public authority. Although this position is an attempt to correct the initial error, its contribution is in any case debatable. The moot point is that it assumes that society at large will bear the cost of rehabilitation, and thus of the building capacity, of some areas that will then be bought and sold in the private sector. To estimate land costs, reference is made to the study by Finpiemonte.[17]

For technical costs, which include both planning and direction of works, a coefficient was used on the construction costs of 5%. For commercial costs, which are needed to cover sales operations, the figure is 2% of sales revenue. General expenses then include all costs in setting up the operation, and these are estimated at 4% of total costs.

If the entire amount of GFS (Gross Floor Space) permitted would be built, that would require setting aside a total of about 129,000 sqm for services that are located partly on the surface and partly underground (circa 30%) and would call for the construction of underground car parks. For these urbanisation costs, reference was made to the figures given in the P.Ri.U., calculated on the basis of municipal tables. Among these, however, the underestimated costs concerning the covering of the railway cuttings have changed. Through figures provided by RCCF-Nodo di Torino, a figure was reached of 2,961,000 lire/sqm.

The pre-urbanisation work (demolition and rehabilitation) and part of the induced urbanisation (the creation of underground car parks) are not included in the analysis. These activities are to be realised by the city, using ministry funding (for a total of 24,904 million lire). The costs of construction were already included in the P.Ri.U.

The sale price for the buildings was calculated by referring to the minimum and maximum property sales prices as surveyed by a national property registry. The values included in the analysis are rather optimistic because of the weight that was attributed to the image of this project and thus to its attractiveness.

All the investment capital needed was considered as borrowed capital. This simplification allows the pure potential of the property operation to be considered. That potential is translated technically into the inclusion of a line of credit in the financial analysis.

The interest rate used for discounting back the cash flow corresponds to the return on zero-risk investments of similar duration: here, the net return of government BTP (Buoni del Tesoro Poliannali) bonds at eight years.[18]

If a private operator was to base its own investment decisions on the results obtained with the Discounted Cash-Flow Analysis conducted here, it would not have the slightest incentive to invest. The NPV (Net Present Value) was, in fact, negative and the IRR (Internal Rate of Return) has a lower value than the interest rate. With values like that, it is clear that the operation shows no profit margin; indeed, it is running at a loss. Such a negative result was obtained despite the most optimistic initial hypothesis: brief duration of the project, the lower cost of the land compared to the theoretical market value, and the financial coverage of a good part of the infrastructural work.

The Discounted Cash-Flow Analysis Applied to Spina 1: Variant of the PRG (January 1998)

In the second analysis of the financial situation in the sector of Spina 1, some figures were changed in the proposal to the variant of the PRG presented at the beginning of 1998.

There are three important variations:

- the plot ratio was lowered (from 0.7 to 0.6 sqm GFS/sqm ST);
- the transfer of the buildable area from one zone to another was eliminated;
- the public areas defined as buildable areas, considering only half of the roads, were reduced.

The reduction in the plot ratio was done for two reasons. First with too high an index, the GFS generated could not be achieved on the relevant land, provoking a mechanism of transfer. Second, such a high coefficient led to a high quantity of standards which, in turn, could find no room on the surface and thus had to be built below ground, with decidedly higher costs.

The elimination of the transfer of area is explained in a similar way. First of all, with the weak demand in the property market – given that the quantity of built works generated in the areas is already considerable and only with difficulty finds operators interested in them – there is no sense in constructing these 'transferred square meters'. The second reason is the reduction provoked in this way in the quantity of the standards to be respected. Similar reasons led to the reduction in the number of public areas considered buildable.

Keeping the same structure of the Discounted Cash-Flow Analysis conducted earlier, these elements were thus changed. In addition, following the new indications, the coverage of the railway tracks was considered the responsibility of the public sector, to be realised through ministry funding. For this reason, that construction work no longer appears in the analysis. When the second application was being conducted, Fiat was negotiating with the city about the possibility of paying a sum instead of meetings the standards.[19] As the city had stated several times that the number of car parks theoretically envisaged by the plan in this zone was effectively overestimated, this would seem to be the most likely outcome.

The results obtained in this second analysis are positive (Table 8.4). The IRR of 12% shows a certain profitability margin for the operator. Although this margin might not repay the investment risk, it is significant in indicating the validity of this channel of revision of the PRG.

If the previous analysis highlighted the reasons why the plan was not feasible, these last results show how the results proposed in January increased the likelihood that the plan would be implemented, even though some problems of the PRG had not been

Table 8.4 Discounted Cash Flow Analysis to Spina 1 (variant of the PRG)

Costs	%	Price (lire)	Quantity	Unit	Total
Real cost suitable areas		560.000	95.859	m² SLP	53.681.040.000
Technical expenses	6				9.131.352.000
General expenses	4				6.087.568.000
Commercial expenses	2				7.206.328.000
Total					76.106.288.000
URBANISATION COSTS					
Construction					
Housing	5				3.559.575.000
Business and personal services	5				298.410.000
Services	10				7.502.950.000
Primary urbanisation					
main sewers		140.000	39.706	m²	5.558.840.000
Network infrastructures		80.000	99.708	m²	7.976.640.000
new roads in the project		80.000	10.234	m²	818.720.000
Artificial standards					
Underground car parks		580.000	9.996	m²	5.797.680.000
Secondary urbanisation					
Square development		250.000	7.669	m²	1.917.250.000
Total					27.632.385.000
CONSTRUCTION COSTS					
Residential building		1.500.000	47.461	m²	71.191.500.000
Business and personal services		1.400.000	4.263	m²	5.968.200.000
Offices		1.700.000	44.135	m²	75.029.500.000

Costs	%	Price (lire)	Quantity	Unit	Total
Total					152.189.200.000
COMPENSATION PLOTS FOR FACILITIES		200.000	26.895	m²	5.379.000.000
TOTAL COSTS					(261.306.873.000)
REVENUES					
Residential building		4.000.000	47.461	m²	189.844.000.000
Offices		4.000.000	44.135	m²	176.540.000.000
Business and personal services		3.700.000	4.263	m²	15.773.100.000
TOTAL REVENUES					382.157.100.000
Interest rate		Annual 5.89%	Six-monthly 2.90%		
NPV					19.936.902.465
IRR ANUAL					12.00%
IRR SIX-MONTHLY					5.83%

completely eliminated. Specifically, even though the GFS had been reduced, the supply of property generated by the plan was still high for a city like Turin. Furthermore, the standards to be provided had fallen, but they continued to represent an important share of costs.

The Proposal for the Planning Variation for the Spina Centrale (September 1998)

Although a statement of intent had been signed in January, the city council had not ratified it. The Spina Centrale had been heavily criticised, because the building density was too high. At present there is no formal measure that modifies the PRG. However, 'guidelines for the revision of the indications of the PRG concerning the Spina Centrale' have been drawn up, based on the following general criteria (Città di Torino, 1998):

- *reduction in the land area*, in particular of the part represented by existing roads, taking the boundary of the zone as the middle of the road, thus cutting the city's building rights;
- attribution of the territorial index of *0.4 sqm GFS/sqm ST* to all areas of the Spina, including all the existing roads while excepting some railway areas. To this index was added a further territorial index for public housing of *0.2 sqm GFS/sqm ST* to be calculated on the built areas, i.e. net of the roads;
- *treatment of the railway areas*: no building rights are attributed to railway areas whose current functions are confirmed (tracks, plant etc.), while the other railway areas, subject to urban transformation, are treated like the other areas of the Spina (0.4 + 0.2 sqm/sqm);
- determination of the areas to be provided for services in two parts: the first part corresponding to the standards for the new constructions planned, the second additional part corresponding to 20% of the ST.

On the basis of these variations, the overall GFS falls to about 40%.[20] The transfers of GFS outside the area in which they are generated are reduced to just 10,000 sqm. The services to be provided also drop considerably (- 35%). They comprise a much higher share compared to the GFS generated, than envisaged by the existing PRG.

The changes proposed go in the direction of the requests made at the start of the year. There was a desire for more green areas and less built-up areas. At the same time, the reduction of the indices and the cut in city spaces make it possible to meet almost all the standards required on the surface, thereby lowering costs for private operators.

The different ways in which these modifications have been reached should, however, be noted. On the one hand, in the course of negotiations, private operators continued to put forward their own requests and to evaluate the city's proposals on the basis of the financial analyses they conducted. On the other hand, the city reached these values 'empirically'. Despite the acknowledged problems found in recent years in implementation, the changes proposed have not been checked through any form of preventive financial assessment.

The choices were made on the basis of political demands, supported at most by checks of a planning nature. It was necessary to lower the amount of built-up area and increase green space as demanded by the left. At the same time, it was important not to reduce the plot ratio too much, so as not to displease the owners. In addition, the 0.6 proposed has been split into two parts, with a share destined expressly for public housing 'for the benefit of the city'. The problem of the standards was also tackled in the same way. The new mechanism that imposes standard of 20% of ST makes it possible, in fact, to maintain a share of services equal to 90% of the standards previously guaranteed by the plan and at the same time to cut considerably the services supplied below ground, with great savings for private operators.

As regards rehabilitation and improvement work, which had also sparked numerous arguments, a 'negotiated' position has been reached. If the areas have a public function, even if they belong to private bodies, then city funds are used. Otherwise, improvement is the responsibility of the owner.

Conclusions

The strongest analogies between Turin and Milan concerning the Crossrail project are visible in the utilisation of the areas used (or to be used). The case of Turin has been amply illustrated. For Milan, it is sufficient to say that the environs of the main stations, Garibaldi and Bovisa, have been designated for the location of particular urban and regional uses. These sites were chosen due to their high regional accessibility. The head office and the new headquarters of the Regional Council are to be built at Garibaldi, the second part of the Polytechnic at Bovisa.

The Crossrail represents, for both cities, an opportunity for large industrial groups to promote their real estate. Most of the land used for the railway infrastructure in fact belongs to large companies. The role of those interested in the transformation of the city is changing. Besides the new interest from the large industrial groups for the part of the project concerning real estate, it should also be noted how strong their contractual power now is with the local authorities.

The Crossrail will obviously help solve the problem of city congestion caused by urban and regional traffic, a matter which can no longer be put off. But it should be noted that the corporate owners of land affected by railway infrastructure plans had already set their budgets according to the (theoretical) value of the high building capacity and the location of significant urban sites. The impression is that the real estate rationale and the consequent repercussions for the budgets have, therefore, ended up co-existing with the original planning expectations and, at times, favour the concrete interests of certain parties.

The large urban areas now in disuse represent property that has been amply amortised in the framework of the various production cycles over the course of fairly long periods of time (sometimes more than a century). Yet, this fact is hardly ever taken into consideration in the evaluations of these areas. These are actually areas that no longer have any value in terms of the way they were used previously. They also have a low exchange value, given that they are located in a situation of excess supply and weak property markets.

It is true that these areas enjoy externalities produced by the city that surrounds them and by the railway infrastructure that concerns them directly. But the externality values are negatively influenced by the extremely high, and above all uncertain, costs of reclaiming the areas. Yet planning policies and tools generally tackle these areas as if they were spaces usable immediately, without restrictions, in dynamic market conditions. Consequently, these areas are assessed at relatively high values. Property value assessment is approached uncritically by planners.

The case study presented here is emblematic in this sense. The new PRG of Turin was conceived merely as a means to regulate property development plans that were thought to be large-scale and certain, thanks to the Crossrail schemes. But this correlation between 'development and infrastructure' has not happened, for various reasons.

The new development plan of Turin, created primarily to control the physical change of the city, must now tackle the financial aspects and economic consequences of transformation. If, originally, these components were not given due consideration (numerous errors were committed in the forecasts), today instead the PRG must in some way adapt itself to the rules of the property market. Compatibility between the plan and the market is necessary because both create constraints and opportunities for the project. In the case of the Turin plan, the fact of not having taken the real situation of the property market into consideration has led to planning choices that turned out to be unfeasible.

One might think, then, that the use of market studies and preventive financial analysis before starting on projects of this kind could have repercussions on planning decisions connected to infrastructures and vice versa. However the case of Turin is an example of under-use and underestimation of these tools. These tools are used only sporadically. When they are used, there is little awareness of their role and almost exclusively by the private sector. It is significant, in fact, that in the process of revision of the PRG of Turin, only the private sector used financial analysis as a tool in the discussions to reach an agreement with the city Meanwhile, the public authority continued to proceed imprudently, basing its decisions mainly on political considerations. In the end the problems encountered in the implementation phase turned out to be mainly of a financial nature. Nonetheless the city did not conduct evaluations of this kind in the process of making variations to the plan.

Notes

1   The 'piano regolatore generale' (PRG) is a masterplan that covers the entire municipality and is drawn up by the municipality itself. The PRG determines the zoning of the local area and the constraints for each area, the transport networks, the areas for public use and facilities of general interest. In effect, it is the most important planning tool in Italy.

2   "Transport investment in urban rail infrastructure is seen as a major instrument in shaping city structure and in promoting economic development. Changes in accessibility resulting from new rail infrastructure should encourage new development around stations. Many cities have invested in new rail systems and some of the associated development has been privately funded. Offices, shops and commercial centres have formed an integral part of the development. As redundant land around railway stations has become available, new development has also taken place. This compact and high density development is a direct result of changes in accessibility and land being released, and much of the recent new prestige office development in the city centre has been of this form" (Banister and Lichfield, 1995).

3   It is believed that the Euro Tunnel, 50 km long, required seven years of work, while work on the Crossrail for the two Italian cities, which are both less than 10 km long, has been going on for more than 15 years and is still not finished.

4   One study has highlighted how BART (Bay Area Transit System) of San Francisco, a rail transport system constructed in the 1970s, had a small but significant positive effect on the price of housing within 1000 feet of the stations, with a variation of 0% to 4% that fell rapidly with the increase in distance from the stations (Blayney Associates, 1978). In no case did the effects of BART extend more than 5000 feet. Burkhardt (1976) and Dornbush (1975) also note a drop in the prices around BART due to some negative effects such as noise and vibrations, the increase in car traffic and the greater accessibility for different social classes and ethnic groups in what were otherwise homogeneous districts.

5   As Wegener (1995) observes: "The evidence is disturbing. It seems to undermine the body of theory expressed by the land-use transport feedback cycle. [...] The apparent dissolution or at least weakening of the interdependency between land use and transport becomes very inconvenient at a time when planners are desperately looking for ways to come to grips with the negative environmental impacts of car traffic in cities under the threat of long-term climate change".

6   The 17/81 Law approves the funding of an integrated programme, with the objective of finding a solution to the most urgent problems of the railways. The programme for the use of the funds provided, amongst other things, for "the procedures for the strengthening of Turin's facilities and the quadrupling of the route from Porta Susa to Torino Stura".

7   This ratio indicates how much GFS ('gross floor space') can be realised for each square metre of land area.

8   "The road and railway networks have always had a decisive role in the structuring of the city, directing its growth and functioning. (...) Today even

more so, they tend to take the form of new urban sites, giving shape to a new landscape and functional infrastructure whose new hidden potential seems to have escaped the notice of many" (Clementi, 1996).

9   Not all the positions on this are in agreement. Edwards (1995) underlines that "Efforts to maximise the contribution to railway profits from property developments at international stations does seem to be a serious threat which the social science community should alert people to. First of all there will tend to be very concentrated over – development at the stations, and the mixed-class populations and small firms who tend to occupy the space round the stations are liable to resist or suffer severe displacement effects – as we know from King's Cross. Preconditions exist for the same sorts of conflict at Zurich, Frankfurt, Brussels, and a number of Italian and other cities".

10   Source: RCCF – The Turin intersection.

11   In Italian planning regulations 'superficie territoriale' (hereafter ST) indicates the entire area of a development plan, not only the built-up section.

12   In Italy the concept of 'standard' in a development plan (introduced by the 1967 law) represented the minimum value for areas destined for services. That value is calculated on the basis of the relationship between the facility already existing or expected within a region and the number of inhabitants to be served by it. The Regional Planning Law for Piedmont provides 25 sqm for inhabitants' services, divided between lower education, facilities of communal interest, green areas, and car parks. The quota for this 'standard' in Lombardy is 26.5 sqm per inhabitant.

13   Source: Milanese Underground Association.

14   The *comparto* is a legal and administrative tool to group ownership. The Municipality Commune can include in its regulations the obligation to create *comparti*, with the possibility of expropriation. This is principally a means of encouraging private owners to implement the indications of the plan. To create a *comparto*, the agreement is needed of owners representing at least 75 per cent of the area as recorded at the *catasto*, the property register.

15   Based on ISTAT figures, IRES has calculated that in 1999 Turin's resident population would be 872,000 people (compared to a million in 1989), of which 30 per cent are people aged over 60 and only 10 per cent of the population are very young.

16   In Italy, the 'standard' (introduced by law in 1967) represents the minimum surface area that must be assigned to services of collective interest in planning regulations, and it is calculated on the basis of the ratio between existing or planned services in a given area and the number of inhabitants who use them. The Piedmont Regional Planning Law specifies 25 sqm of services per inhabitant, divided into areas for education, green areas, car parks and services of collective interest.

17   Through the parameters provided, a price for the 'suitable area' was set at 560,000 £/sqm GFS.

18   'Buoni del Tesororo Poliannali' like governments bonds with fixed rate of intent. The rate used here, of 5.89 per cent per annum, was the performance of the BTPs (government bonds considered to be of zero risk) at the time of the first of the analyses presented here.

19   This is the part of the standard over the legally demanded percentage and the 10 per cent of the 'superficie territoriale'.
20   The overall GFS falls from 1,992,500 sqm to 1,138,500 sqm.

References

Banister D., Lichfield N. (1995), Introduction, in D. Banister (Ed.) *Transport and Urban Development*, E & FN SPON, London, pp. 1-16.

Blarney Associates (1978), *The Study of Property Prices and Rents: BART Impact Study*, Metropolitan Transportation Commission, Berkeley.

Burkhardt, R. (1976), *Summary of Research: Joint Development Study*, Administration and Managerial Research Association, New York.

Cervero, B. and Landis, J. (1995), Development Impacts of Urban Transport: A US Perspective, in D. Banister (Ed.) *Transport and Urban Development*, E & FN SPON, London, pp. 136-156.

Chicco, P. and Saccomani, S. (1993), Torino: un sogno immobiliare contro il declino?, in F. Indovina, *La città occasionale*, pp. 151-222. Milano: Franco Angeli.

Città di Torino (1996), *Programma di riqualificazione urbana Spina.1* - ambito 12.9 PRG, November.

Città di Torino, Divisione Urbanistica (1998), *Indirizzi programmatici per la revisione delle indicazioni di PRG relative alla Spina Centrale*, September.

Clementi, A. (1996), Nuovi modi di intendere gli spazi infrastrutturali, in A. Clementi (Ed.) *Infrastrutture e piani urbanistici*, Quaderni Blu n.4, Fratelli Palombi Editori, Roma, pp. 247-262.

Dornbush, D. (1975), BART - Induced Changes in Property Value and Rents, in *Land Use and Urban Development Projects, Phase I, BART Impact Study*, Department of Transportation and U.S. Department of Housing and Urban Development, Washington DC: U.S.

Edwards M. (1995), Critical Issues in Regional Rail Investment, in D. Banister (Ed.) *Transport and Urban Development*, E & FN SPON, London, pp. 127-137.

Fareri, P. (1990), La progettazione del governo a Milano: nuovi attori per la metropoli matura, in AA.VV, *Metropoli per progetti*, Il Mulino, Bologna, pp. 163-217.

Finpiemonte (1996), *Programmi di Riqualificazione Urbana* - Rimodulazione economico-finanziaria: criteri e metodologia utilizzata, Consulenza per il Comune di Torino.

Garelli, M. (1993), Pianificazione e progetti a Torino, in F. Indovina, *La città occasionale*, Franco Angeli, Milano, pp 227-271.

Grieco, M. B. (1994), *The Impact of Transport Investment Projects upon the Inner City: A Literature Review*, Aldershot, Avebury.

Indovina, F. (1993), *La città occasionale*, Franco Angeli, Milanon.

Nomisma (1996), Osservatorio sul mercato immobiliare. *Rapporto quaadrimestrale Luglio 1996, 2-96*, anno IX.

Prizzon, F. (1995), *Gli investimenti immobiliari*, Celid, Torino.

Pugno, R. (1994), Costi relativi alla realizzazione del sistema Passante del Servizio Ferroviario Regionale (SFR), in IReR, *Il servizio ferroviario regionale. Una strategia di investimento*, Stampa IreR, Milano, pp. 9-36.

Secchi, B. (1996), Un'interpretazione delle fasi più recenti dello sviluppo italiano, in A. Clementi (Ed.), Infrastrutture e piani urbanistici, *Quaderni Blu n.4*, Palombi Editori Fratelli,pp. 27-36.

Wegener, M. (1995), Accessibility and Development Impacts, in D. Banister, *Transport and Urban Development*, E & FN SPON, London, pp. 157-161.

Appendix

Discounted Cash-Flow Analysis

Discounted Cash-Flow Analysis is a quantitative economic evaluation technique. It is one the forms of financial evaluations that analyse the current net investment result in relation to a project.

The structure of the *Discounted Cash-Flow Analysis* consists in a matrix where the incoming and outgoing financial flows are on the lines, and the columns show the periods of time in which the project is divided, a duration which is set arbitrarily according to the time that it is estimated the project will take.

The criterion on which the evaluation of the profitability of a project is based consists in the discounting back of cash flows. Two synthetic indicators of financial profitability are obtained: the *Net Present Value* (NPV) which represents the discounted back sum of the cash flows, and the *Internal Rate of Return* (IRR), which is the interest rate determined by the return on the capital invested.

Calculation of the return is obtained by considering for each time period the revenues, costs, and finally the balance. The flows thus obtained must be discounted back, as it is not possible to compare the balances from different periods. They cannot simply be added up, as there is compound interest, so they are discounted back through an appropriate coefficient that allows the future financial value of an investment to be discounted back in order to obtain its current value.

Discounting back is calculated through the formula:

$$C_0 = C_n \; \frac{1}{(1 + i)^n}$$

where C_0 = initial capital (capital at time zero)
$\quad\;\; C_n$ = capital at time n
$\quad\;\; i$ = interest rate

The sum of all the discounted back capital flows (F), which constitutes the NPV (*Net Present Value*) and thus the initial value, is given by:

$$NPV = (\sum_{t=1}^{n} F_t \, (1 + i)^{-t}$$

where NPV = net discounted back value
F_t = capital flows at time t (with t a variable between 1 and n)
i = interest rate

The NPV is the first indicator of the profitability of the operation. If it is positive, this means that the investment has produced a net benefit (even if this is not enough to establish the threshold of the acceptability of the investment); if in contrast the NPV is negative, this means that the discounted back costs are higher than profits, and the operation is thus inadvisable.

As the discounting back coefficient is inversely proportional to the interest rate and time, the weight of costs will be greater the more concentrated they are in the early periods and the higher the *i* are. The function of the NPV thus varies according to *I*, falling when this increases.

The NPV is a monetary amount that does not consider the capital invested, and it is thus necessary to use a second profitability indicator, the *Internal Rate of Return* (IRR). The IRR is the value of *i* that annuls the NPV, i.e. the rate that renders equivalent the negative and positive flows of a property operation. The IRR is calculated for repeated cycles, calculating the NPV with gradually rising (or falling) discount rates until its value approaches zero.

For an operation to be acceptable, the IRR must be higher than a minimum threshold defined previously by the operator, a threshold that can be interpreted as the sum of three distinct components:

- the share necessary to compensate for (or neutralise) forecast inflation;
- the share necessary to compensate an investment at zero (or almost zero) risk;
- the share necessary to reward investment risk (Prizzon, 1995).

The first share depends on the forecasting of the mid-term financial markets, the second is the return on a zero risk investment, for example government bonds. On the operational level, the return that covers these two elements is determined by comparison with government bonds of the same duration. It is instead difficult to determine the minimum acceptable level of IRR that can provide a return for the investment risk, and this is why in recent years analysis techniques have been developed in this direction (sensitivity, probability and risk analysis).

9 Congested Evening Rush Hour Traffic in Antwerp: a Sustainable and Integrated Policy Approach

ANN VERHETSEL

Introduction

This case presents a tool for transportation planning and makes some preliminary calculations for the urban region of Antwerp (Belgium).[1] The tool and the process of developing it are based the notions of sustainable urban development and integrated urban governance. In this report, the focus is on sustainability, which is examined by evaluating solutions to traffic congestion during the evening rush hour. The concept of accessibility is at the centre of the model.

Not only does the Antwerp urban region include the most important industrial area in Belgium, but this area is also linked to the Antwerp harbour, which plays a major role in distribution networks throughout Europe and the world. The accessibility of the harbour area is crucial to the economic performance not only of Antwerp, but also of Flanders and Belgium as a whole. The fierce competition with Rotterdam (only 100 km away) and the other harbours in the range from Antwerp to Le Havre, mainly in the domain of container traffic, makes the congestion on highways a strategic problem. That problem calls for an effective solution in the short run as well as in the long run.

During the past decade, the distance between place of work and place of residence has increased continuously. Since both the average- and higher-income groups have suburbanised to such an extent, and since economic activities continued to be located mainly in the central area, there is a high degree of 'wasteful commuting' (Hamilton, 1989). This poses a threat to the economic performance of the urban area, but also to the quality of life. Moreover, non-urbanised environments have become very scarce in the surroundings of Antwerp, so the segregation processes also have a number of irreversible ecological effects. These trends are not acceptable, and alternative scenarios inspired by sustainability are needed. The present model compares the effects of two alternative approaches: a trend scenario versus a sustainable one.

Various authorities concerned with problems of mobility in Antwerp have organised themselves into a temporal network for the development of a multi-modal traffic model. Among them are the Ministry of the Flemish Community (departments of infrastructure, of spatial planning and of housing), the City of Antwerp, the national railway company, the Flemish company for public transport (buses and trams), as well as several regional and local economic authorities. The project was managed by the Mobility Cell of the Flemish administration, starting from the stage of looking for partners who wanted to invest and continuing until the organisation of the effective use of the developed model was ready. This cell itself was created at the beginning of the nineties. It brought people together from different departments of the Flemish administration to develop an integrated policy for mobility issues. So the classic hierarchical and functional relations between departments and policy levels were circumvented in this project. It was carried out by a group of consultants working in a temporary association, though they also hired a number of academic research groups. So the project drew upon wide-ranging expertise on the mobility problem in Antwerp.

The model that was developed distinguishes four phases: trip generation, transport distribution, route choice and testing out the model against reality. The concept of accessibility is made operational in the stages of transport distribution and route choice. The number of trips from a certain zone to various destinations is calculated. Simultaneously, a modal split is made: public transport, car or bicycle. The distribution and modal split are made to coincide by means of a gravity model. According to this model, the number of trips between two traffic zones depends on the number of departures, the number of arrivals, and the force of attraction between the zones. The force of attraction is determined by means of distribution functions per mode of transport, per degree of urbanisation, per motive for travelling and per car availability. In the case of motor traffic, intersection delays and congestion are taken into account. The route choice is determined by means of travel resistance, expressed in terms of time and costs. Travel resistance is a weighted combination of travel time and travel costs. Travel costs are expressed in terms of time and are subsequently added to the travel time. In a number of areas, the travel resistance against trips by car also takes into account the time spent finding a place to park and the costs of parking.

The relative accessibility is measured by means of the distance (in time and money) from one location to another. It is a place accessibility measure that fits into the 'narrow sense' interpretation of accessibility (section: 'The multi-modal Traffic Modal and the Policy Measures').

We first give an overview of the policy measures and then show how these are introduced in the model by presenting two alternative scenarios. Next we describe the resulting traffic flows for each alternative scenario as a whole. Subsequently, we analyse the relative impact of each policy package on evening rush hour movements. Finally, we focus on the impact of some specific planning and infrastructure measures. On this issue some empirical evidence from the USA and Western Europe has already been brought together by Banister (1995). Within each analysis, we focus on the results for traffic in the inner city.

The Multi-modal Traffic Modal and the Policy Measures

This traffic model describes private and public road transport as well as haulage. In the current analysis, we concentrate on policy measures that can provide a solution for passenger traffic problems. Freight transport is included in the model, but the studied measures have little impact on haulage. The analysis focuses on the busiest period of the evening rush hour, i.e. between 4:30 and 5:30 p.m. The model was initially developed for the Antwerp region (Figure 9.1: Network of roads included and area studied, and figure 9.2: Area studied (origin-destination traffic zones) and road network). In the future, it should become possible to make similar calculations for other Flemish regions and for Flanders as a whole.

Source: Verhetsel, 1999

Figure 9.1 Network of roads included and area studied

The passenger transport model is a simultaneous demand model that makes integrated calculations of relational patterns and modal choice on the basis of time and costs. The model was inspired by the so-called Randstad model (Rijkswaterstaat, 1988). More information on the model can be found in Verhetsel (1998).

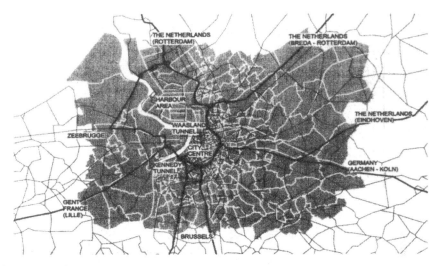

Source: Verhetsel, 1999

Figure 9.2 Area studied (origin-destination traffic zones) **and road network**

After the model had been tested and modified for 1991, packages of policy measures were drawn up and their impact on future (2010) traffic flows were measured. Packages of measures may relate to four distinct aspects:

- planning: the geographical distribution of residents, the working population, and employment per traffic zone in the year 2010;
- infrastructure: adaptation of the basic network for motor traffic and public transport;
- regulations: changes due to legislation on parking, vehicle occupancy, etc.;
- financial aspects: variations in petrol prices, parking fees, public transport fees, road pricing, etc.

Two alternative scenarios were investigated:

- a trend scenario that gives an indication of how the mobility issue will develop if the current policy is maintained. This means

Table 9.1   Overview of packages of measures for the two scenarios

Trend scenario	Policy scenario
Planning: The emphasis is on a continuation of current spatial developments: further de-urbanisation, on the one hand, and continuing growth of the tertiary sector coupled with a greater concentration of employment in inner city areas and a number of peripheral locations. These trends are guided by the possibilities created by the 'Gewestplannen', i.e. regional development plans designed in the period of growth in the 1960s and 70s. These plans encompass the judicial framework that exists today.	*Planning:* A strict location policy, concentrating future growth (in housing and employment) in and around cities. Spatial organisation in accordance with the 'Ontwerp Ruimtelijk Structuurplan Vlaanderen'. This plan provides a Flemish context for the replacement of existing land-use plans.
Infrastructure: Modifications to the motor traffic network required for the sake of maintenance and safety; completion of current projects of regional importance; strategic plan for a Flemish-European transport axis. An unchanged public transport policy is expected to reduce service as a result of increasing costs and road congestion.	*Infrastructure:* No fundamental investments in the motorway network; funding only for safety and maintenance projects and for current projects of regional importance. Public transport is highly developed throughout the region, with good infrastructure and frequent service. All residential nuclei in rural areas are served. Public transport is given absolute priority in urban areas.
Regulations: Status quo 1991.	*Regulations:* An extensive electronic traffic control system ('telematics'), separate carriageways for buses and car-poolers, closing of approach roads; cars of non-esidents are banned from inner cities, implementation of a strict parking policy in areas of employment.

Trend scenario	Policy scenario
Financial measures:	*Financial measures:*
Introduction of a motorway tax for heavy lorries, and a tax on carbon dioxide emissions included in the price of petrol and diesel. A decrease in expenditure on fuel in real terms; public transport fares follow inflation; higher parking fees.	Users of individual transport are charged for congestion and external costs by means of a new motorway tax on all vehicles; road pricing which will see variable car costs increase by 50% inside urban ring roads; a carbon dioxide tax. Car ownership is discouraged by abolishing tax incentives for the purchase and use of a car for commuting.

Source: Verhetsel, 1999

that demand and supply develop according to market principles that do not have to include external costs. In fact, this scenario supports the actual power relations in the mobility debate (priority for automobilemanufacturers, infrastructure developers, etc.);

- a policy scenario that aims at managing mobility by controlling transport demand through spatial planning based on proximity and compactness, and by making optimal use of public transport to satisfy mobility requirements. This scenario tries to incorporate a set of measures that fit within a policy of sustainable development.

Table 9.1 provides an overview of how the various packages of measures might be implemented for the two scenarios. This overview was put together with the help of a great many experts in the field. The measures are a reflection of the most likely course of action taken if future policy in one or the other direction were to be outlined at present.

Impact of Alternative Scenarios

Table 9.2 provides an overview of the shifts in traffic generation during the evening rush hour. The calculations are based on the implementation of the trend and policy scenario. The car kilometres are converted into passenger kilometres by multiplying them by the average vehicle occupancy (1991: 1.26; trend scenario: 1.28; policy scenario: 1.56). One notes that in 2010, i.e. the planning horizon, the resulting traffic flows are quite different for each of the scenarios. A continuation of the current policy trend would result in a growth of the number of passenger kilometres during the evening rush hour by some 38%. The corresponding number of additional car kilometres would amount to 43%, compared to 11% for public transport. In absolute terms, we are talking about an increase of over two million kilometres during the evening rush hour. The additional

Table 9.2 Impact of the different scenarios (generated passenger kilometres during the evening rush hour)

	1991	Trend scenario			Policy scenario		
	Absolute	Absolute	Difference with 91	Index 1991 = 100	Absolute	Difference with 91	Index 1991 = 100
AREA STUDIED							
Public transport	806 578	892 314	85 736	111	1 703 797	897 219	211
Car	4 679 464	6 703 827	2 024 364	143	4 196 322	-483 142	90
Total	5 486 042	7 596 141	2 110 100	138	5 900 119	414 077	108
INNER CITY							
Public transport	21 914	20 531	-1 383	94	41 130	19 219	188
Car	45 170	48 509	3 340	107	35 746	-9 424	79
Total	67 084	69 040	1 957	103	76 876	9 792	115

Market shares of public transport/cars (%)

	1991		Trend scenario		Policy scenario	
	Pt	Car	Pt	Car	Pt	Car
AREA STUDIES	15	85	12	88	29	71
INNER CITY	33	67	30	70	54	46

Source: Verhetsel, 1999

traffic burden that this would place on available infrastructure is hardly imaginable. The traffic density per carriageway would increase enormously, going beyond capacity at certain crucial places. The capacity breach is most dramatic in the trend scenario on the Antwerp ring road. On stretches of road where capacity is exceeded, the explanatory power of the model is limited. In particular, queuing on carriageways leading to the Kennedy tunnel (in both directions, to France and to The Netherlands), the Waasland tunnel and the approach to the motorway to Germany would lead to behavioural changes. Alternative means of transport, roads and timing would become very interesting. Under the policy scenario, growth in the number of passenger kilometres during the evening rush hour would be limited to 8% (+414,077 km), with public transport accounting for the entire increase (+111%, +897,219 km).

The effects in Antwerp's inner city (figure 9.3: Inner city and road network) are slightly different from the average developments

Source: Verhetsel, 1999

Figure 9.3 Inner city and road network

in the area studied as a whole. In the trend scenario, traffic in the town centre would increase by only 3%. The growth would be accounted for in its entirety by car use (+7%), while use of public transport would decline (-6%). The policy scenario would result in more passenger kilometres in the evening rush hour (+15%, +9,792 km) than current trends seem to suggest. This is due to a greater concentration of people working and living in the town. As a result of this proximity, however, a great deal of travelling between home and work could be done on an adequately developed public transport system (+88%, +19,219 km). Cars would be left in the stable (-21%, -9,424 km). In the next section we identify the measures that are responsible for the different impact of the scenarios.

The Packages of Measures Assessed

In order to measure the sensitivity of the traffic model to the various packages of measures, we have made calculations for a number of fictitious scenarios. In this particular study, we successively incorporated into the reference scenario of 1991 first the package of planning measures from the trend scenario and afterwards the policy scenario. The resulting forecasts are entirely fictitious, for the reason that within one scenario there must be a certain logic connecting the various packages of measures. It is, for example, rather absurd to concentrate socio-economic activities for the year 2010 geographically in urban areas, without assuming that the public transport network would be adapted accordingly. In other words, these are fictitious forecasts that are merely intended to measure the sensitivity of the model to packages of policy measures.

We present the results of the fictitious forecasts with, in succession, the effects of the different policy measures relating to planning, infrastructure, regulatory and financial aspects. In this order, the measures are determined increasingly less by 'spatial' characteristics.

When comparing the various packages of measures with each other (Table 9.3), it appears that the more 'spatial' the measures, the smaller the impact on the kilometres generated. The impact of the spatial component (planning and infrastructure) is clearly secondary to the other components: financial measures in particular are very effective. Rodier and Johnston (1997) summarise some studies in the USA with similar results: road pricing strategies reduce daily vehicle miles by 10-15%, land-use measures might project a 10%

reduction but such measures take up to 20 years to become truly effective.

Table 9.3 Effects on traffic generation of the packages of measures (growth indexes 1991-2010)

		Trend scenario		Policy scenario	
		Area studied	*Inner city*	*Area studied*	*Inner city*
Planning	Public transport	106	99	105	98
	Car	110	97	106	99
Infrastructure	Public transport	101	103	115	130
	Car	100	100	99	97
Regulatory	Public transport			118	134
	Car			95	80
Financial	Public transport	110	100	152	140
	Car	127	108	62	78

Source: Verhetsel, 1999

The measures taken in the trend scenario have little impact on traffic flows in the inner city. Only a continuation of current financial trends would result in an increase of car use in the inner city (+8%), though not to the same extent as in the area studied as a whole (+27%).

The policy scenario, by contrast, does yield strongly divergent results for the area studied and the inner city. The planning choices concerning proximity and urban compaction produce just a fractionally smaller (-1%) traffic generation. In actual fact, it concerns a status quo. The infrastructure and regulatory measures have a much bigger impact in the inner city than in the area studied as a whole. The use of public transport relative to car use increases to an unexpectedly high level. Financial policy options have the greatest impact, though this is less pronounced for inner cities than for the area studied as a whole. Together with the regulatory context, they result in a split of the number of kilometres travelled between public transport and cars in the proportion of 45/55. With respect to a sustainable development of mobility, this is much more desirable than the proportion of 33/67 generated by the trend scenario.

The results for Antwerp's inner city show a fairly strong deviation. This implies that certain local effects should not be underestimated. But neither of the packages of planning measures leads to fundamental change; results of the other packages are much more pronounced. Larger towns that wish to address the mobility problem should not expect too much from planning measures. Infrastructure measures may stimulate people to make use of public transport. But especially regulatory and financial interventions, which are usually directed from higher policy-making levels, implicitly have a stronger effect. This is an important motivation for local authorities to keep a close check on higher authorities and to keep reminding them of the geographically differentiated effects of general policy. Of course, it is also possible for local authorities to take more regulatory and financial measures than has been the case till now.

Planning measures are the least effective. Infrastructure measures and regulatory measures have more effect, while financial measures clearly have the most significant impact. The varying impact is more pronounced in the policy scenario than in the trend scenario. This is quite logical, as current policy is largely maintained in the trend scenario so that the differences with respect to 1991 are much smaller. Our conclusion is identical to Wegener's (1995, pp. 160): "This implies that in metropolitan areas with inexpensive transport, little planning control and a deregulated land market, policies to influence location or travel behaviour only by incentives must fail". We fully agree with his explanation, which distinguishes between accessibility as a scarce resource (as is the case in classic urban economic theories) and ubiquitous accessibility. In our study area – with subsidised car trips from home to work, without serious time losses up till now and without external costs to be paid – we work under the assumption of ubiquitous accessibility. Then location changes have little effect on traffic patterns, and vice versa.

Analysis of the traffic model demonstrates that planning and infrastructure measures have little impact because behaviour is changed only among a small group located in or nearby the traffic zones subject to those measures. In contrast, regulations and especially financial measures affect nearly everyone involved in the evening rush hour. The detailed analysis of planning and infrastructure measures presented in the next two sections will make this clear.

A Detailed Look at Planning Measures

The planning measures consist of a different geographical distribution of residents, working population and employment per traffic zone. The trend scenario emphasises a continuation of current (1991) developments. Towards 2010, the distribution undergoes two changes: further de-urbanisation; and the continuing growth of the tertiary sector together with a greater concentration of employment in inner-city areas and a number of peripheral locations, as foreseen by the actual regional development plans. For the working population, the trend scenario shows absolute growth in the suburban area. In the policy scenario, there is also a process of suburbanisation of the working population, but it is more modest. In the model that predicts the evening rush, more people will have the suburban traffic zones as a destination in the trend scenario compared with the policy scenario. The maps depicting the evolution of employment in the traffic zones show similar distribution patterns, whereby the inner city loses employment to traffic zones immediately outside the inner city. Although some important local differences can be found in the two scenarios, in the model there will be no big differences in the number of employees originating from various traffic zones during the evening rush hour.

Table 9.4 provides an overview of the generated passenger kilometres by car and by public transport. The figures refer to i) the year 1991, ii) 1991 with the package of planning measures in the trend scenario, and iii) 1991 with the package of planning measures in the policy scenario. So the purely hypothetical question that we are trying to answer here may be stated as follows: if the infrastructure network, the regulations, and the financial context all remain unchanged in comparison to 1991, while working population and employment change to the 2010 level, then what are the implications in terms of mobility increase and distribution if either spatial diffusion (trend scenario) or spatial compactness (policy scenario) are the prevailing planning concepts?

Table 9.4 Effects of planning measures (generated kilometres during the evening rush hour)

	1991	1991 +Trend			1991 +Policy		
	Absolute	Absolute	Difference with 1991	Index 1991=100	Absolute	Difference with 1991	Index 1991=100
AREA STUDIED							
Public Transport	806 578	855 460	48 882	106	843 485	36 907	105
Car	4 679 464	5 165 265	485 802	110	4 981 975	302 511	106
Total	5 486 042	6 020 725	534 684	110	5 825 460	339 418	106
INNER CITY							
Public Transport	21 914	21 607	-307	99	21 554	-360	98
Car	45 170	43 886	-1 284	97	44 860	-310	99
Total	67 084	65 493	-1 591	98	66 414	-670	99

Market share of public transport/car (%)

	1991		1991 + Trend		1991 + Policy	
	PT	Car	Pt	Car	Pt	Car
Area studied	15	85	14	86	14	86
Inner city	33	67	33	67	32	68

Source: Verhetsel, 1999

Both scenarios for the future generate additional kilometres in the area studied compared to 1991. This is quite normal, as the prognosis for the year 2010 anticipates an increase in the number of residents, the working population and employment. The growth in absolute figures is slightly smaller for all kinds of traffic in the policy scenario. This, too, is logical: in view of the greater proximity of residential and working areas, the total number of kilometres travelled is restricted.

In the trend scenario, the increase in generated kilometres is mostly accounted for by cars (+485,802 km). The absolute increase of the distance travelled during the evening rush hour by public transport (+48,882 km) is only a tenth of the additional kilometres travelled by car.

The fact that the trend scenario generates a greater increase in distance travelled by public transport (+48,882 km) than the policy scenario does (+36,907 km) is due to two developments. First, the place of residence and the place of employment are concentrated more strongly in the latter scenario. Secondly, the public transport network is not being adapted to the spatial characteristics. The impact of the infrastructure context is analysed separately in the following section. The resulting market shares in the area studied for passenger travel by car or by public transport are hardly influenced at all by planning measures: +1% for public transport in the policy scenario.

Analysis of Infrastructural Measures

The Impact of the Infrastructure Policy Package as a Whole

What is measured here is the hypothetical impact of the provision of various new infrastructure projects if planning of the division of living and working remains unchanged and if regulations and the financial context are held constant (Table 9.5).

The total number of generated kilometres in the area studied during the evening rush hour is more or less the same for the two scenarios (+1%). The infrastructure measures in the trend scenario are minimal. Consequently, so is the impact on driving and public

Table 9.5 Effects of infrastructure measures (generated kilometres during the evening rush hour)

	1991	1991 +Trend			1991 +Policy		
	Absolute	Absolute	Difference with 1991	Index 1991=100	Absolute	Difference with 1991	Index 1991=100
AREA STUDIED							
Public Transport	806 578	813 469	6 891	101	928 989	122 411	115
Car	4 679 464	4 696 513	17 049	100	4 620 047	-59 417	99
Total	5 486 042	5 509 982	23 940	100	5 549 036	62 994	101
INNER CITY							
Public Transport	21 914	22 482	568	103	28 516	6 602	130
Car	45 170	45 092	-78	100	43 820	-1 350	97
Total	67 084	67 534	490	101	72 336	5 253	108

Market share of public transport/car (%)

	1991		1991 +Trend		1991 +Policy	
	PT	Car	PT	Car	PT	Car
AREA STUDIED	15	85	15	85	17	83
INNER CITY	33	67	33	67	39	61

Source: Verhetsel, 1999

transport use. In the policy scenario, significant investments are made in public transport. As a result, one would see a number of car users (-1%, -59,417 km) changes over to public transport (+15%, +122,411 km). This implies that the respective shares of car and public transport in passenger transport during the evening rush hour do not change in the trend scenario, while there is a slight impact in the policy scenario: +2% for public transport.

In the inner city, the infrastructure measures have a much bigger impact than in the area studied as a whole. The use of public transport relative to car use increases to 39/61. With respect to a sustainable development of mobility, this is much more desirable than the proportion of 33/67 generated by the trend scenario.

As mentioned above infrastructure measures have a rather local impact on the evening rush hour behaviour of commuters. Only people using the links where infrastructure will be changed or living/working in the neighbourhood of these changes will respond to these investments. Some other cases, like the impact of the Supertram in Sheffield, show similar results (Lawless and Dabinett, 1995). In the next section we give some examples.

The Impact of some Projects

The Department of Infrastructure of the Flemish Ministry is, of course, very interested in getting some idea of the effects of their planned projects. The projects presented here are solutions for the evening rush hour congestion in Antwerp, a problem that local and regional policy-makers bring up frequently. In the local and regional media, we often find discussions with arguments for and against particular measures, but up till now devoid of facts. For the government, these results could help promote rational problem-solving and policy-making. The analysis presented here was drawn from internal reports of the Flemish Department of Infrastructure. The work was done by Verhaert and Vanuytven (1997). The results are obtained from running the trend scenario for 2010, whereby traffic densities with and without the new projects are compared.

Project 1: closing existing ring road (A) + new second ring road (B+C)
The new link A is a very valuable alternative for link A', the latter being a highly congested part of the existing ring road (Figure 9.4:

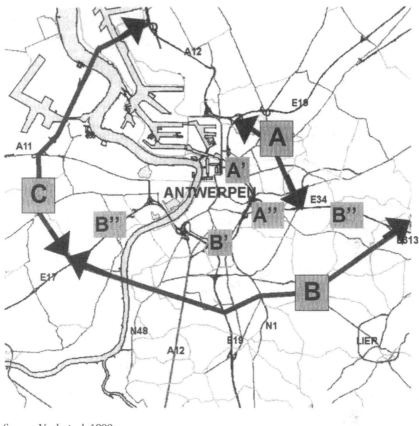

Source: Verhetsel, 1999

**Figure 9.4 Project 1: closing existing ring road (A) + new second
ring road (B+C)**

Project 1: closing existing ring road (A) + new second ring road
(B+C)). According to the trend scenario, the new link would take
over 2500 passenger car units (pcu) per hour. The new link also
positively affects the congested link A'', leading to a decrease of up
to 4500 pcu/hour! The new link takes 6000 cars and 400 lorries
during one evening rush hour.

Link B (E34→E17) shares traffic with the rest of the existing ring
road R1 (link B'). The new link over the river (Schelde) would take
2500 cars/hour away from the existing Kennedy tunnel. The

overcrowded parts of the E17 (B′′) and E34 (B′′′) would also benefit greatly from this new link.

Link C creates a new highway between the E17 and the northern tunnel in the middle of the harbour (Liefkenshoektunnel). This toll tunnel is actually not used in an optimal way. One reason is that the link between the E17 and A11 is non-existent. Moreover, being the only toll tunnel in Flanders, people are not accustomed (or willing?) to pay out of pocket to use such infrastructure. People with interests in the harbour consider this new link vital to the growth and success of the harbour. On the other hand, environmental groups protest against the link because of the valuable polder landscape that would be disturbed. This new link would hardly attract cars out of the Kennedy tunnel. Only the local roads between E17 and A11 would benefit (a decrease of 500 cars/hour). Thus the strategic importance of the link for the harbour does not concur with a solution for the congested Kennedy tunnel and the existing ring road (R1).

Project 2: creation of a new link over the River Schelde north of Antwerp This project is an alternative solution for the problems in the Kennedy tunnel and on the existing R1 ring road as a whole (Figure 9.5: Project 2: Creation of a new link over the River Schelde north of Antwerp city). The new link would take 5000 cars/hour and 1000 lorries/hour during evening rush hour. This results in about 70, 000 pcu/day, which would still leave 150, 000 pcu/day for the Kennedy tunnel. Without the new northern link, the Kennedy tunnel would take 190, 000 pcu/day. The R1 loses 3000 pcu/hour, which means that some bottlenecks on the R1 would not be solved by this alternative. The old Imalsotunnel loses 1600 cars/hour and the toll tunnel would lose an extra 600 cars/hour.

Comparing the results of project 1 and 2, it is clear that the northern link over the Schelde would take 1500 cars and 800 lorries more than the southern link. The southern link would take traffic mainly from the Kennedy tunnel. Besides cars, the northern link takes more lorries and also relieves the other tunnels. The fact that project 2 closes the ring road R1 can improve the management of the traffic system (e.g. by telematics). But the access roads from and to Antwerp remain overcrowded! To address that problem, public transport systems (park and drive) and financial measures to limit private transport would be necessary.

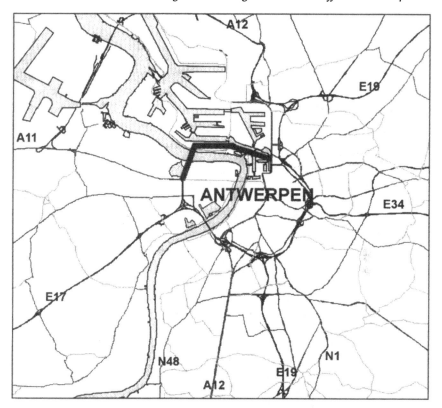

Source: Verhetsel, 1999

Figure 9.5 Project 2: creation of a new link over the river Schelde north of Antwerp city

Comparing the two projects sheds light on the options. First of all, we see that a simple investment in one or a few links cannot solve the problems in the traffic system. Secondly, we see that a political decision is needed to set priorities about which problems need to be solved first with scarce means.

Conclusion

A traffic model to forecast the multi-modal passenger flows during the evening peak in 2010 is used to analyse the impact of different policy measures to reduce traffic problems in the Antwerp region. In fact, by introducing different policy packages and isolated policy measures in the model, we estimate the sensitivity of the model to these inputs. The model was built very carefully by experienced model-builders, and the calibration for 1991 gave good results, except for links that reach saturation level. Thus it can give us an idea of the impact of different traffic policies.

A package of policy measures that follows the actual trend results in totally different passenger flows and modal choices than a package designed in accordance with a policy scenario that aims a sustainable development. Differences emerge when comparing the impact of planning, infrastructure, regulatory and financial measures. We cannot expect too much from planning measures. Infrastructure measures may stimulate people to make use of public transport. Nevertheless, regulatory and financial interventions have stronger effects.

Planning measures will influence the behaviour of only a few people who travel in the evening rush hour. Under the given circumstances, few commuters will be able to decrease the distance between work and home. The fact that the Antwerp area already has a very extensive suburban zone solution for the evening rush hour, even in the long run.

Major investments in infrastructure can result in solutions for the evening peak traffic on some links. But even these costly and politically difficult projects only give partial and local results. In most cases, additional regulatory and financial measures are needed.

From the point of view of planners and geographers, more research should be done on the local effects of different measures, giving special attention to the implicit spatial effects of regulatory and financial policies. We should not lose sight of the fact that all the analyses presented here refer to the Antwerp area, a region with a strong, dominant centre and a dispersed suburban zone. New analyses are required for the rest of Flanders, with the exception of Brussels.

This case illustrates how a tool for transportation planning can be used in the discourse on sustainable urban development. Nowadays, the Flemish Mobility Cell and the Flemish company for public transport (de Lijn), play a central role in the network. They exploit the model, make further investments in personnel to keep the model running, and provide (for a fee to non-partners) calculations for public and private organisations.

Notes

1   The study was conducted under the authority of the Flemish Government with the help of the Administration of Infrastructure. We thank ir. E. Peetermans for contributing to its conceptual development and ir. C. Verhaert and ir. H. Van Uytven for offering practical assistance. This chapter is reprinted from the Journal of Transport Geography, vol. 9, Ann Verhetsel. The impact of planning and infrastructure measures on rush hour congestion in Antwerp, Belgium, pp. 111-123, copyright 2001 with permission from Elsevier Science.

References

Banister, D. (1995), *Transport and urban development*, F&FN SPON, London.

Hamilton, B. (1989), Wasteful commuting again, *Journal of Political Economy*, 97(6), pp. 1497-1504.

Lawless, P. and Dabinett, G. (1995), Urban regeneration and transport investment: a research agenda, *Environment and Planning A*, 27, pp. 1029-1048.

Rijkswaterstaat (1988), *Het Randstadmodel. Verkeer en vervoer in de Randstad. Technische rapportage*, Dienst Verkeerskunde – DHV, Den Haag.

Rodier, C. and Johnston, R. (1997), Incentives for local governments to implement travel demand management measures, *Transportation Research A*, 34(4), pp. 295-308.

Verhaert, C. and Vanuytven, H. (1996), *Sensitiviteitsanalyses en onderzoek van beleidsmaatregelen met het multimodaal verkeersmodel*. Ministerie van de Vlaamse Gemeenschap, Departement Leefmilieu en Infrastructuur, Multimodaal Verkeersmodel Antwerpen.

Verhetsel, A. (1998), The impact of spatial versus economic measures in an urban transportation plan, *Computers, Environment and Urban Systems (CEUS)*, 22(6), pp. 541-555.

Wegener, M. (1995), Accessibility and development impacts, in D. Banister (Ed.) *Transport and urban development*, F&FN SPON, London, pp. 157-161.

10 Development of Central Public Spaces in Geneva: Changing Uses and Images

VÉRONIQUE STEIN

Introduction

Today, every European city is facing the challenge of developing its public spaces. They have to decide what do with disused industrial sites and how to redistribute populations within the city. Thus the authorities engage in discussion with the city's inhabitants on which perspective to adopt and which kind of contact to establish for each of the projects.

In this contribution, we consider how this challenge is met in Switzerland, a country that prides itself on its frequent use of direct democracy. That approach entails very specific rules for actors' participation and a greater general commitment by local residents.

Geneva's situation illustrates the actions carried out by authorities in a gentrification process. It also illustrates the diversity of actors involved in the projects, the complexity of their relationships, their scope of power, and so on. We show that fragmentation (on the level of the projects, the ideas and the decision-making venues) is characteristic of public space management in Geneva. As a result, things are too often handled as they come up. This lack of a global vision can lead to unexpected and rather drastic consequences.

For the Geneva case, we propose to analyse the modifications in the behaviour of certain groups among the population when new infrastructures are implemented.[1] This was developed by authors like Rapoport (1986) for whom the use and development of public spaces exemplify human-environment relations. We must ask ourselves to what extent the built environment affects human behaviour and well-being in urban areas.

We argue that *top-down* development schemes cannot be successful by themselves. They have to be supplemented by *bottom-up* approaches that include all parties from the urban environment (particularly local residents and different social groups) and must take into account the different geographical levels (such as districts, communes, conurbation) (Bailly and Scariati, 1999; Decoutère, 1996). Indeed, new ways of initiating public space planning processes are emerging (co-operation, forms of partnerships) which bring about big changes in users' practices and images.

Our case is supported by the results of two surveys. These were conducted to assess the role of central public spaces[2] through the eyes of the inhabitants of Geneva. For a number of reasons, we used the transformation of Geneva's Coulouvrenière district to illustrate our point. These three reasons are elaborated below.

First of all, the Coulouvrenière case enables us to examine the notion of *the city centre*. Indeed, the way inhabitants think about central public spaces and their facilities will determine the value of the city centre. Is the city centre perceived in terms of advantages and disadvantages? Is it, for the people who live there, easily accessible? What kinds of transportation are available? Have recent efforts to improve the city centre been fruitful?

Secondly, we want to illustrate *certain divisions* that come up when urban projects are being designed and implemented. We believe these diverging views are a source of tension and even deadlocks. These divisions are particularly visible in the creation and management of central public spaces. With this in mind, we try to determine if the authorities in charge of Geneva's public space management have a common policy. If so, we try to ascertain whether this policy is guided by the will to improve the inhabitants' quality of life or, on the contrary, to merely make the city more attractive.

The question then becomes whether the Coulouvrenière public spaces were designed as a city showcase with an international perspective or whether it was created for and by its inhabitants as a 'real' place to live (and in this last case, which part of the population it was specifically designed to serve).

Thirdly, the Coulouvrenière case suggests that it is not enough for a development project to be *aesthetically or technically perfect*. For its inhabitants to embrace it, a project must offer a sense of belonging. Practices cannot be concocted from scratch, and all projects contain unexpected elements (mainly linked to behaviour, the users' values, etc.). Although the management and creation of public spaces in Geneva has largely been taken over by certain groups of architects and town planners, it is not as easy for those who live and work there to take part in these projects.

Managing Public Space

Inhabitants at the Heart of the Debate

A study of public spaces cannot be carried out without first looking at the public that uses them: long-standing resident-inhabitants, daily users and commuters. While public spaces such as streets, squares and parks are open and accessible to all-in theory, it's often quite different in reality.

Public spaces offer a number of variations. The way in which people make a space their own depends on the location, the time of the day, the season, the person's age and lifestyle. We develop all sorts of ties with the public spaces in our neighbourhood, city or region, combining practices and image. Although we use these spaces in our daily lives, the question remains: how strong is our desire or our ability to act upon these? And to what extent are we qualified to make decisions regarding the future of these spaces as citizens, inhabitants or users?

While these questions are not new, they revive the debate on the role played by inhabitants in decisions concerning town and country planning. What are the roles and responsibilities of inhabitants/users compared to those of professionals or elected representatives?

These themes are actually quite complex when it comes to central public spaces. While it is relatively easy to determine the identity of the inhabitants of specific areas or districts, this is not the case for those of the city centre. Who should really be taken into account in central public space development projects? Should it be those who live near the central spaces? Or should it be the population of the city as a whole? How do users and businesses fit into the equation?

With this in mind, we think it is important to go beyond the functional level and take into consideration people's images. Although many sites are rarely visited in practice (central public spaces in particular), they have a strong symbolic value (Rémy, 1991).[3] This leads us to postulate that even if we are not users or inhabitants of these spaces, we should still have a say in how they are managed.

Function and Development of Public Spaces

Most people would agree that the key functions of public spaces are circulation and meeting/sociability. Yet few would agree on which aspects to change to improve the quality of life, encourage varied uses of the city centre, and ultimately make the city a more pleasant place to live.

Planning and analysis of public spaces often centres around traffic issues. Thus, leaving aside other aspects inherent to public spaces, the debate focuses on the search for a balance between the various means of transportation and optimum accessibility (Bonanomi, 1996). It is important to remember at this stage that the way infrastructures are designed depends closely on the underlying views of the developers in charge.

Are they concerned about aesthetic qualities? Do they put an emphasis on street furniture such as benches? Do they influence mainly outdoor public spaces or do they include the buildings which border these areas as well? Do they believe that a 'participative' approach is essential?

We believe that it is crucial to understand the perspectives taken by specialists. While public spaces are used for traffic and meeting places, they are also intended for expression (cultural, for example) and special events. They foster links between different parts of the city and act as territorial markers or landmarks, both visually and symbolically (Goffman, 1973). The various urban facilities must be evaluated in order to review which are 'best' suited to the many facets of public spaces and to determine which groups of the population are most served in each situation.

Town Planning and Dividing Lines

Our hypothesis is that when an urban development project is proposed, the diverging views mentioned below can crystallise into tension and eventually hold up to some extent the management process under way.

Let us start by distinguishing four main groups with various points of views and different stakes in any project: regional authorities, city centre authorities, citizen groups (associations), inhabitants/users. Between these actors, the following dividing lines may be discerned:

- the first division lies among the ruling authorities. It is an inter-institutional conflict setting the *city centre* against the *region*;
- secondly, within the previous example, we find *diverging views*. these are of two different kinds:
  - above all, there is a discrepancy between the views of *architects* and *planners*, who are concerned mainly with the design of the project, and the *engineers* for whom the technical data is most important;
  - certain tensions may also arise between actors with a predominantly *aesthetic* or *morphological* view, versus other actors with a stronger *participative* tendency.

The first view (morphological) is that of actors, or groups of actors, who wish to work on the infrastructure level of projects, focusing mainly on physical transformation. From this perspective, aesthetic values are predominant and public space is very much regarded as scenery that can be adapted. In addition, enhancing public spaces is considered an 'asset' for the city in its competition with other urban areas and therefore constitutes an option for urban marketing.

The second tendency (social) refers to actors who make it a priority to work on the social level (in a broad sense) and for whom the residents' quality of life is essential;

- thirdly, tensions can also result from views and interests of the *authorities*, on the one hand, and those of *users/inhabitants* on the other;
- a last line of division exists between the ruling authorities who favour a project's *realisation* as such and those for whom the *participation* of users/inhabitants in the decision process is essential.

Production and Management of Public Spaces in Geneva

Context: the Complexity of Structures

Looking at Geneva only as a small international city (an area of 1,600 hectares with a population of 176,486), we tend to forget the complexity of its political and administrative structures.

Let us recall that the Canton of Geneva is divided into 45 communes having a surface area of approximately 246 square kilometres and a population of nearly 400,000 inhabitants. The City of Geneva is the largest commune in terms of population and jobs. In terms of its influence zone, Geneva and its surrounding area have 600,000 inhabitants who are spread out over two Swiss cantons and into neighbouring France.[4]

Even local residents themselves very often misunderstand the various levels of authority, especially the distinction between the *commune* and the *canton*. Indeed, Geneva's case is peculiar to Switzerland: although most of the planning decisions depend on the canton authorities, the communes act as experts and have important prerogatives. They publish notices about all documents related to the management of their territory, yet they are still under the canton's supervision. Thus, commune autonomy is not automatic and remains an active request, especially for the City of Geneva. In 1984, city authorities officially formalised a Town Planning Department (which had already existed for a few decades), giving proof to this problem. The present situation is the result of the 1931 constitutional law, which divided up the competencies between city and canton. Since then, the city is responsible for the public domain and its equipment (schools, culture, etc.), whereas canton authorities have decision-making power for most other planning fields.

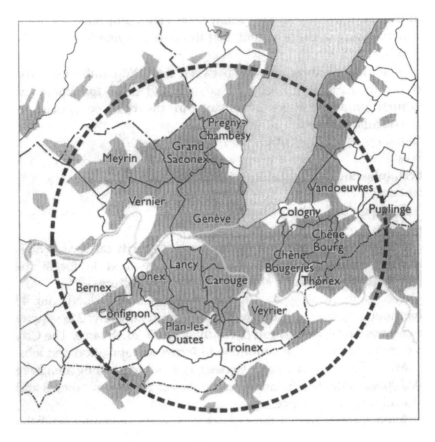

Figure 10.1   Geneva's conurbation

A Desire to Improve the City Centre

In Geneva, as in other European cities, the city centre has implemented a series of strategies in the last few years aiming to improve its attractiveness, both locally (in regard to the surrounding area) as well as internationally (in its competition with other cities).

In order to do this, the canton authorities have proposed, since 1996, to revise the *Plan Directeur Cantonal* of 1989. One of the main purposes of this new plan is to assess the quality of life in the city by looking particularly at public spaces, mainly squares and parks, which make up the major 'breathing places' in a dense city like Geneva.

A real will to improve links between public and natural settings by creating a viable network of streets (mainly for pedestrians) is underway. Meanwhile, an effort is being made to identify and highlight the main poles of activity and social life.

From this perspective, the question of transportation remains fundamental. It is closely connected to the issue of accessibility (both private and public) and of public spaces management. It brings us back to the very essence of what 'urban' means (Joye and Kaufmann, 1998).

The question of how the Coulouvrenière projects fit in with the general will is discussed later in this paper.

The Stakeholders

The main actors involved in *public space management* in Geneva may be classified as follows:

- *Municipal departments.* The city owns 400 hectares of public domain (one quarter of the city's territory), which entitles it to a leading role in the decision-making process. This explains why public space planning and development is and always has been an area in which the city authorities have a particularly strong desire to gain more power.

  Among the services involved are the Town Planning Department (*Service d'Urbanisme*), the Urban Development Department (*Service d'Aménagement Urbain*), the Public Parks Department (*Service des Espaces Verts*), and the Engineering Department (*Service des Constructions et de la Voirie*);
- *Canton departments.* These authorities are mainly in charge of planning and organising the public transportation system for the entire Canton of Geneva.

Depending on the situation, other groups feel deeply involved with the development of central public spaces. Among others, these include neighbourhood associations, lobbies (automobile and economic, for example), and political parties.

The Main Difficulties

A few remarks should be made about the context we have just explained:

- there is a *competition* between and a *multiplicity* of administrations or departments which deal with public spaces, implying many views and showing a certain lack of coherence. A difficulty to come up with a single path to follow is apparent. The people in charge of the departments involved often play two roles: one dealing with (city) management, the other one being political. If they happen to be involved in both of these areas, they usually operate on different levels, that of the city and the canton. On top of this, there is the occasional thirst for power and possible conflicts of interest. As a result, projects often become the business of one single person. Accordingly, the procedures for getting projects off the ground are often burdensome (in time, energy spent, etc.).

  In this context, the debate over transportation issues in Geneva is particularly heated. Although the canton is theoretically responsible for *transportation systems* and *policies*, tensions between the canton and the city can be noticed. Indeed, the city centre has shown some specific and acute problems that require specially adapted answers. This is why municipal authorities currently play an active role in the elaboration of public transportation policies. The position of the city authorities is to favour modal report in order to free the public domain from private traffic and to improve access to and from neighbourhoods. To do this, the city welcomes new transportation opportunities in order to enhance the public domain for pedestrians and cyclists. The canton has more or less accepted this view as long as it does not overly hinder existing automobile traffic;

- certain *well-known and respected architects* in Geneva seem to have taken over the development of public spaces. As a result, a specific view is imposed, while other groups of actors are excluded from the process. This seems particularly troubling, in that there is an obvious lack of freedom surrounding the inventiveness and actions taken in this domain;

- the complexity and proliferation of planning instruments and services has brought about a lack of communication and transparency between *users* and *planners*. Realising this problem, municipal services have focused on encouraging inhabitants to participate in the decision-making process;

    Various departments recommend integrating users more. Yet their views remain strongly divided, resulting in a certain confusion over the definition of *participation*. Some departments feel they should listen but that in the end it is up to the architect/town planner to propose and finally approve the project. Some think they should carry out public surveys during the design stage and reformulate it if necessary, while others feel it is enough to explain things once the project has been implemented;

- certain tensions arise between the *inhabitants* and *authorities* regarding public space management. The population of Geneva appears to demand a greater level of participation in the decision-making process. At the same time, the populace voices its opposition to *top-down* projects.[5]

The Case of the Coulouvrenière

The following sections highlight the previously mentioned difficulties that came up the most during the projects to revitalise the Coulouvrenière district.

Situation and Evolution

The Coulouvrenière district is located in the extension of the city centre's main thoroughfares, luxury shops and banking sector. It is currently an important subject of debate, but its purpose has not yet been clearly established.

We defined an area of study within this district. We narrowed it down to a square surrounded by cultural buildings, a footbridge linking the two banks of the river, a sector of the banks of the Rhône, and a site belonging to the Municipal Engineering Department (*Service Industriels*) that was recently allocated to subcultural groups.

When we speak of the Coulouvrenière, we essentially mean this area.

Let us take a quick look at the numerous transformations that the Coulouvrenière has undergone over the last few decades. For a long time, this district was reserved for market gardening and military exercises. Washboats docked on the Rhône River made genuine meeting places for women and public baths livened up the banks of the river, providing a relaxed atmosphere.

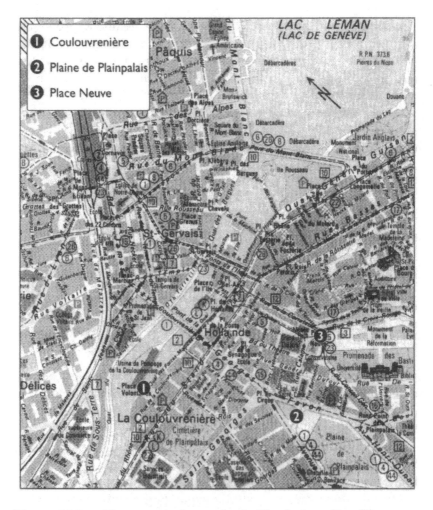

Figure 10.2   The area under analysis (Coulouvrenière (1), Plainpalais (2) and Neuve (3))[6]

It was only at the end of the 19th century that this district became the heart of Geneva's industrial sector. The Coulouvrenière has been the symbol of industry and working-class housing for three generations.

Today, cultural activities (mainly alternative) dominate in this area. The users tend to cause public nuisance (noise, drugs)[7] that partially drive away the property sector. A number of disused industrial buildings have thus been taken over by subcultural groups, ranging from illegal tenancies to more or less long term agreements with the local authorities. Moreover, a number of associations have set up their headquarters in the Coulouvrenière. Today, this district plays a major role for a number of groups in precarious situations: it includes a centre for the homeless and a training centre for school dropouts.

Most recently, more prestigious cultural activities have been introduced in certain buildings, with a tendency to transform the district's users and reputation.

Projects and Achievements

In the 1970s, many projects came to life in the Coulouvrenière area, launched mainly by Geneva's alternative and cultural circles. Although they were already preoccupied with the district's future at the time, the city and canton authorities finally implemented a series of development projects[8] in the 1990s. These projects, some of which have been completed or are currently under way, fall within the scope of modern urban problems. Let us go over them:

- a section of the Coulouvrenière covers the *Rhône riverbanks;* these banks are largely underused for leisure or walking. They also tend to be overlooked as a functional corridor providing quick access to the city centre on foot or by bicycle. Efforts have been made to improve the general accessibility of the area as well as the continuity of the network. These measures were carried out as part of the *Fil du Rhône* project, relating to a series of small, successive areas along the river. This project acts as a town planning guide allotted to a number of actors (private/public). The fulfilment of its various stages depends on their willingness to co-operate.

The project was launched by the Municipal Decoration Fund, although various city and canton planning departments participated in the project's development. Recommendations were also requested from other commissions, such as the commission in charge of heritage preservation. One year of negotiation (1994) was necessary before work could start on the first construction sites (1995-96).

The project was awarded to an architect who called in artists to join him. The original aspect of this development project was to integrate architects and artists from the beginning during the design stage. In this context, works of art are not simply displayed here and there in various urban places, as it is usually the case. Rather, artists participate directly in the 'improvement' actions and choose the appropriate locations where they will work later on.

This idea of *continuity or fluidity* for pedestrians was also worked on by another city service[9] for the whole of Geneva. The *Pedestrian Plan*, as it is known, is aimed at improving acces to the city centre through technical works (pedestrian paths, sidewalks) and information campaigns. Improving the users' general sense of security is also one of the project's main goals. The project is part of an overall reconfiguration of the transportation system and is carried out from the perspective of sustainable development.

As we said before, the canton authorities are in charge of transportation policies, but city authorities have actively contributed through preliminary analyses and sugestions. Among them, the city puts particular emphasis on improving the public domain for pedestrians.

In this specific project, the Town Planning Department makes an inventory of the demands of the various actors involved and presents them as *needs* before entering the operational phase. For the Coulouvrenière area, the Town Planning Department has done a considerable amount of work in contacting neighbourhood associations and setting up a negotiation process;

- as previously mentioned, the role of Coulouvrenière as a cultural district, an essentially alternative one, has dominated the area for a long time. However, the creation of a real *alternative* centre raises a number of concerns among the authorities and the district's inhabitants (represented by an association). Should this cultural hub be reinforced and considered an essential part of Geneva? Or, on the contrary, would it be better to encourage a mix of activities and populations?

In other words, should this district be developed for a specific group or for the district's inhabitants as a whole? Finally, how will these developments affect Geneva's image and reputation?

The answer to these questions reflects two prevailing perspectives. The first emanates from the alternative circles themselves, strongly backed by the Left. Their argument is that the city does not have a *viable* project for this district while the demand for artistic and cultural premises abound and the city centre includes a large area of empty commercial and industrial spaces (a total of 300,000 square metres).

The city and canton authorities have another viewpoint. Although they had previously tolerated (even accepted) the convergence of this alternative culture in the district[10] under certain conditions, today they seem to be changing their minds.

A new tendency is now emerging favouring high-class culture along with urban improvement policies. A disused building has been rehabilitated and devoted to more prestigious cultural activities; a very popular café has been transformed into a luxurious pub; and the façade of an alternative culture building has been restored. Finally, a little square, which up to now was an area for pedestrians and cyclists during the day and a meeting spot during the night, has become a car park.

Another argument recently put forth by the city authorities and the district's inhabitants, who are irritated by the presence of the alternative movement, involves defending the need for green spaces in such a densely populated district. Along the same lines, the idea of a central square (not necessarily from a spatial standpoint) that could serve as a social meeting place for the inhabitants is defended in the Coulouvrenière. This district currently lacks such a space, yet it is home to a large number of foreigners who could benefit from it;

- the Coulouvrenière area comprises a series of disused industrial buildings that many consider an *industrial heritage* that is worth saving. Beyond their aesthetic and architectural value, these buildings constitute important anchoring points and help make up the collective identity of Geneva. They are reminiscent of a distinctive period with its set of social and symbolic values, activities, styles, etc. In relation to public spaces, these buildings border the squares and streets, reinforcing or transforming both the uses and images.

While many people from the Coulouvrenière agree that these formerly industrial buildings need to be preserved, the proposals on how to do so vary. It seems that the authorities, particularly the ones in charge of preserving the heritage, have recently put a lot of effort into renovations to modify the image and reputation of the area, long considered as 'squalid', and to create incentives for social diversity.

Alternative (arts) groups who occupy the disused industrial buildings are more interested in the allocation and use of these spaces for cultural events, professional studios, etc. For them, these buildings provide large interior spaces that can be easily transformed. Moreover, the buildings are relatively centrally located.

Finally, for a minority of the population of Geneva, these buildings are a legacy of another age worthy of preserving. This attachment may not necessarily concern the best-known buildings, such as important factories. Rather, it may apply to certain isolated little places, such as an old popular café or an abandoned house covered with ivy. Generally, it's at the moment transformations have been decided upon, the petitions are drawn up by the district's or the city's residents to stop any development.

Project Evaluation

This section aims to 'test' – from the point of view of inhabitants/users – the value placed on the Coulouvrenière district and to measure the role of Geneva's central public spaces in general.[11]

The users'/inhabitants' *interest* in the various projects was measured through questions in our survey. A sizeable portion of Geneva's population included in our survey (38%, survey B) *does not know*, or *never goes* to the Coulouvrenière, compared to four and 6% for the two other areas (Plainpalais and Place Neuve).

Asked about the *importance for the city of Geneva* that users attribute to the Coulouvrenière, only 36% answered positively compared to 6% and 72% for Plainpalais and Place Neuve.

The *degree of appreciation* for the area was assessed through certain qualifiers: 21% of users questioned (survey B) qualified the area as negative (including the elements 'ugly/unsightly'), compared to 5% and 28% for Place Neuve and Plainpalais, respectively.

The *visiting frequency* of the Coulouvrenière is low: only one quarter of those polled say they go there regularly (survey A). This is clearly limited to a specific social category, i.e. the most cultural types such as students, people in the professions, and managers. Besides, 45% of users polled (survey B) stated that they come from outside the district, a relatively high proportion compared to 40% and 70% (Plainpalais and Place Neuve). This last result goes to show how much the selection of users is not so much a question of distance as it is of social background. Hence, the district seems to be losing its neighbourhood (or popular) quality in favour of a cultural district.

Moreover, it is even more interesting, that the differences between public spaces are waning now that people are frequenting the places in question. The *sense of social mix* is raised by the question: *Would you say that particular groups or everybody uses this place?* (survey A). The respondents who do not frequent the places feel that they are very much intended for specific population groups (alternative movements in the case of the Coulouvrenière). Consequently, those respondents do not have a very high opinion of them (and vice versa).

But if the three public spaces under analysis are appreciated differently, *the city centre as a whole* is widely valued. 84% of the respondents think it is *very or somewhat pleasant* to spend some time in the centre.

In addition, for 66% of the respondents, the central public spaces constitute a very important aspect of *the quality of life* in Geneva.[12]

The value attributed to recent *developments* was also assessed by the survey. First, 65% of users find that the Coulouvrenière area *has changed a lot during these last few years.* This proportion is rather high when compared with the other two areas (10% and 34%), which are considered relatively stable (survey B). This shows that the changes resulting from the different development projects do not go unnoticed in general.

This relative openness to the developments was confirmed by a second question: *What is most characteristic of this area, according to you, giving it a special atmosphere?* When the users were asked this question, the leading answer (for 47% of users polled) was *morphological* aspects (such as buildings, facilities, developments, natural elements), to the detriment (for 29% of users polled) of *social* aspects (type of people, activities). This attention to physical elements has also been demonstrated by previous studies (Joye, 1992). The result is particularly striking for our area of analysis, which is very rich culturally speaking.

The users do not think that the *accessibility* of the Coulouvrenière, which was improved by the various development projects, is particularly good. In fact, only 61% of the users perceive this area as being accessible, compared to 88 and 96% for the other areas (survey B).

Finally, the elements that *users value the most and should not disappear* are above all the natural ones. This is not surprising in a district as dense as the Coulouvrenière. The four trees bordering the small square were expressly mentioned by some of the users polled. An appreciation of the natural environment was also expressed for other parts of the city. Moreover, this general tendency was already noted in previous studies in Geneva, which happens to be the most densely populated city in Switzerland (Compagnon and Kaufmann, 1994; Bishop, 1986).

Certain *former industrial buildings* which have been renovated were also mentioned as places that should be preserved. For the Coulouvrenière area, this reflects both an attachment to the uses and the architectural/aesthetic value of the buildings. In contrast, the most disturbing element mentioned (and thus eligible to be destroyed) was a very modern building.

Finally, we tried to assess the *nature of participation requested* by the population of Geneva for public space development projects. In our sample, the question *'According to you, who should participate, along with the authorities, in the important decisions regarding public spaces located in the city?'* (answers: *users of these spaces, inhabitants of nearby districts, of the city, of the canton, associations, political parties*) shows that a large proportion of population (89%) is willing to have the users, as well as the inhabitants, take part in the management process (survey A). This percentage decreases when it comes to associations and political parties. That result emphasises the importance of participants' proximity.

These results could be refined by a question such as *In which situation would you be willing to start a petition or public demonstration?* The inhabitants of the Coulouvrenière are particularly willing to invest in pedestrian pathways (67% of the inhabitants polled compared to 45 and 55% in other areas) and are against the destruction of historical buildings.

These *results taken together* lead to the following reflections:

- the efforts of the authorities to promote greater social diversity through cultural activities in the Coulouvrenière area and to improve the image of the district have not had positive results yet. If we look at both practices and images, the population of Geneva has not, in general, embraced the project;
- the alternative circles are the most 'visible' in the Coulouvrenière area. This means that the major actors generally focus on this specific group and consider it as the only type of user of the district. Apparently, there have been no studies examining the kinds of actions that would be suitable for other users or inhabitants of the district (such as elderly persons, fringe groups, etc.);
- while most users agree that the area is in the process of being transformed, efforts to improve accessibility have gone largely unnoticed. This is demonstrated by the fact that most inhabitants feel motivated to carry out an action in order to improve the connections between the district and the city centre;
- in general, Geneva's population does not really care about the Coulouvrenière. Nonetheless, the users and inhabitants of the district feel a sense of belonging, which is reflected in their attachment to their local heritage.

Conclusion

An exhaustive evaluation of the projects covered in this chapter would go far beyond the scope of our study. Besides, these projects are not yet completed. Nonetheless, we shall consider certain tendencies and impressions already mentioned, focusing on the results that we just described.

First of all, it should be remembered that the land development projects of the Coulouvrenière are clearly in line with the *city centre reconstruction process* that kicked off in the early 1990s in Geneva. The purpose of this process, let us not forget, was to 'balance out', to some extent, the general tendency to develop the city peripheries.

The city evidently seems to have identified the Coulouvrenière as an important urban pole that should be improved and integrated into the rest of the surrounding area. The authorities have attempted to encourage *social mix* and to bring together the various actors involved in the public space management processes. However, as with any gentrification process, the reality is that specific social groups end up being favoured through the creation of certain facilities.

During the whole time these projects are being implemented, we can sense some difficulty in bringing the people of Geneva into the process. The projects are 'perfect' on paper and in the blueprints, but turning them into reality is another story altogether. For example, the Fil du Rhône project is brilliant from the standpoint of its design and the actors it brought together (municipal services, architects and artists). Nevertheless, we are uncertain as to the scope of the debate it has sparked. While the project is well understood and has been integrated by the actors involved, it remains cut off from certain inhabitants/users. Is this a result of the presentation of the finished product only (and for a specific public, mainly the elite)?

Similarly, let us point out that the *Fil du Rhône* project has made it possible to examine *the role of art in public spaces*. In this case, art is an integral part of the city. *The Fil du Rhône project* constitutes a test case that can be reproduced elsewhere in the city. However, more thought should be given to the relationship between inhabitants and artists. Artists often feel that the population in general is not aware of cultural and artistic values.

If we go back to the *diverging views* that we brought up in section 'Town Planning and dividing lines', the morphological/aesthetic approach is clearly more dominant than social considerations. This makes us wonder if the city's projects for the Coulouvrenière are intended to truly improve the quality of life in the city. It appears to us that they are intended more for creating a kind of shop window for showing off the city's outer image than for improving the well-being of its residents.

Looking *at the city/canton problems* that we have outlined in section 'The Main Difficulties', the handling of the Coulouvrenière area has been relatively simple. The various city departments were the main actors in the projects. We should note, however, that few links between the various projects have been established up to now (for example between the *Fil du Rhône* and the *Pedestrian Plan* projects) and that no *general action plan* has been developed. Moreover, we can see a lack of foresight concerning the links between developing outside public spaces and adjacent buildings (formerly industrial buildings). The various stages seem to have been carried out as they came along and according to the given circumstances (financing, political will, etc.).

The Geneva case resembles the other Swiss case in this book, *the Zurich West Project*, in terms of industrial site revitalisation and efforts initiated by local authorities to improve the area's reputation and attract new populations to the centre. The two areas also show some important contrasts. The Zurich West Basin is a non-residential area, where it was easier to put in new facilities and launch a negotiation process. On the contrary, the Geneva case was characterised for a long time by working-class housing and is now a very densely built-up district. Also, the scale of the two areas is very different.

In this chapter, we have tried to demonstrate that, particularly in the case of the development of public spaces, it is essential for users/inhabitants to see concrete and tangible improvements and derive substantial advantages from the projects proposed by the authorities.

We therefore believe that the critical point of a project is its being embraced by the various actors involved and in particular by the inhabitants/users. Thus, during a project's design phase, an insufficient consideration of symbolic and social elements can seriously weaken its success.

In conclusion, we believe that fragmentation dominates the public space management process in Geneva. The problems are not really addressed on the right scale, and each actor handles its issues without crossing over into other domains. The inevitable result is a 'lame' consensus and the situation whereby finished projects don't really relate to each other.

Notes

1   We specifically refer to transportation (in the broad sense) and leisure/cultural infrastructures. For a typology of infrastructures, see chapter 1 of this book.
2   These surveys are part of the Cost-Civitas research 'Public space and urban dynamics' (Institute for Research on the Built Environment, Lausanne; Dept. of Geography, Geneva ; 1998-99).
3   We also refer back to the distinction between places *of life* and *known* places developed by P. Amphoux et al., 1988.
4   These numbers are based on the 1990 Census.
5   This has been proven many times through the latest local elections.
6   The two other areas that are compared with the Coulouvrenière district (see section 'Project evaluation' of this chapter) are shown on the same map. These two areas are located in the very centre of Geneva and are nodes for public transportation. One is *Place Neuve*, a very prestigious *square* surrounded by monumental buildings and characterised by uncontested heritage value. The other one, *Plaine de Plainpalais* can be defined mainly by its multiplicity of uses (market, circus, etc.). Although the Coulouvrenière district is in the city centre, it is the most peripheral and the less prestigious of the areas.
7   Let us point out that these problems remain quite minor compared to those in other major cities and that they are handled relatively well in Geneva. Certain groups of actors (we come back to this later) have nonetheless highlighted these nuisances.
8   Although a local district plan was drafted in 1970, it was not implemented.
9   The Town Planning Department (Service d'Urbanisme) of the City of Geneva leads this project.
10 Among other things, this concentration made it possible to control certain social groups better.
11 Through these two surveys (see footnote 2), the evaluation (positive, negative, etc.) was carried out on the level of the *users* and of the *inhabitants* of the canton (who do not necessarily come to this area).
    - A survey (which we call 'A') of 900 households residing in the Canton of Geneva was conducted in November 1998. Samples were selected in order to verify the impact of spatial and social contexts on public space evaluation.

This survey was conducted by telephone (each call lasting approximately 20 minutes) on the basis of a questionnaire including mostly closed questions. The questionnaire aimed to test if there was a link between, on the one hand, social/spatial position and insertion, local participation, attachment to neighbourhood, lifestyles, leisure activities (public/private), involvement in the various urban debates and on the other hand, practices and values related to public spaces.

- A survey (which we call 'B') of 555 users of the three central public spaces of our analysis was completed 'in the field' during the month of September 1998. This survey was carried out at various times and consisted of a short questionnaire (five minutes) of mostly closed questions. For the Coulouvrenière area, the questioning was done at the square itself and along the bridge. The results from the two polling spots were lumped together · (average) for figuring the results of this survey. The goal of this survey was to identify key issues related to activities and preferences, mobility and accessibility, effective and relative (subjective) social mixing/cohesion.

[12] These last two results are independent of the respondents' socio-professional status, age, sex and nationality.

References

Amphoux, P. (Ed.) (1988), *Mémoire collective et urbanisation*, IREC/EPFL, Lausanne, CREPU, Genève.

Bailly, A. and Scariati, R. (1999), *Voyage en géographie*, Economica, Paris.

Bishop, J. (1986), *The best of two worlds*, Bristol.

Bonanomi, L. (1996), Pour un urbanisme de proximité, in C. Jaccoud, M. Schuler, M. Bassand, *Raisons et déraisons de la ville*, Presses Polytechniques et universitaires romandes, Lausanne, pp. 359-376.

Compagnon, A. and Kaufmann, V. (1994), *La ville mal-aimé*, Volet logement, Univox, IREC-EPFL, Lausanne.

Decoutère, S. (1996), Finalités et modalités du management territorial, in S. Decoutère, J. Ruegg, and D. Joye (Eds.), *Le management territorial*, Presses Polytechniques et universitaires romandes, Lausanne, pp. 25-38.

Goffman, E. (1973), *Mise en scène de la vie quotidienne*, Minuit, Paris.

Joye, D. (1992), Habitants des quartiers, citoyens de la ville, *Rapport de recherche n° 98 pour le PNR Villes et transport*, 25, IREC/EPFL, Lausanne.

Joye, D. and Kaufmann, V. (1998), 50 ans d'aménagement du territoire à Genève, *Les annales de la recherche urbaine*, 80-81, pp.93-99.

Rapoport, A. (1986), The use and design of open space in urban neighbourhoods, in D. Frick (Ed.), *The Quality of urban life*, De Gruyter, Berlin, New York, pp. 159-175.

Rémy. J. (1991), Espaces publics et complexité du social, *Espaces et Sociétés*, 62-63, pp. 5-7.

11 The Zurich West Development Project

PETER GÜLLER AND WALTER SCHENKEL

Introduction

The subject of this chapter, namely the negotiation process for the up-grading of Zurich West should be seen from the perspective of governance and network management. The theory behind this perspective is described in chapters 1 and 5. The study presented here is based on the assumption that urban governance is much more about managing complex problems and actor networks than about making decisions and enforcing legislation.

This study is an attempt to bridge the gap between physical planning and the concrete decision-making process that precedes development. The Zurich West Project (Figure 11.1) must be seen as an element of four *planning levels*.

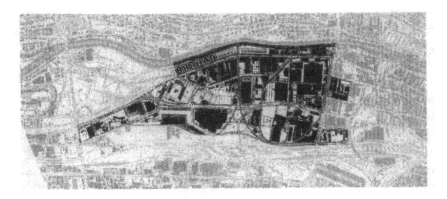

Figure 11.1   Zurich West: the site

I. the general framework is formed by basic political, social, and economic conditions such as federal and cantonal legislation and socio-economic developments of the City of Zurich;

II. a second dimension consists of urban developments and large projects influencing the process in Zurich West;

III. there is the Zurich West Development Project itself, which has emerged out of the *Stadtforum*, and the Impulse Group for Up-grading Zurich West (IGA);

IV. microstructures such as individual enterprises, local urban quality, and local history form the forth dimension; landowners also pursue their own planning activities.

The *industrial district of Zurich* is not simply the area where manufacturing took place in the last century. It contains housing facilities related to industrial employment as well as vocational schools related to production. The whole area extends over three kilometres from the Central Railway Station. The industrial site in the narrower sense (Zurich West Development Project) comprises about one sq. km between the Limmat River and the main railway corridor. The general de-industrialisation process, which has been occurring in developed countries since the 1970s, sent the area into decline. It brought an increasing concentration of foreign migrants and socially disadvantaged groups, strong dissent regarding land-use policy and heavy traffic (highway construction, logistic enterprises). The area is now about to be converted into a broader spectrum of urban functions: residential, cultural, leisure, office, services, high-tech industry manufacturing, breeders for enterprises (Figure 11.2). Till recently, it was not an address for more wealthy households. Yet this is changing, with implants of lofts and other modern types of urban housing. And while part of the old industrial plant has disappeared, some very specialised high-tech niche production has remained. Furthermore, various up-grading efforts are already under way, initiated by individual industries and local authorities.

A few years ago, the Mayor of Zurich created the *Zurich Stadtforum*, a platform for various interest groups. Its mission was to discuss the problems faced in Zurich West and find ways to solve

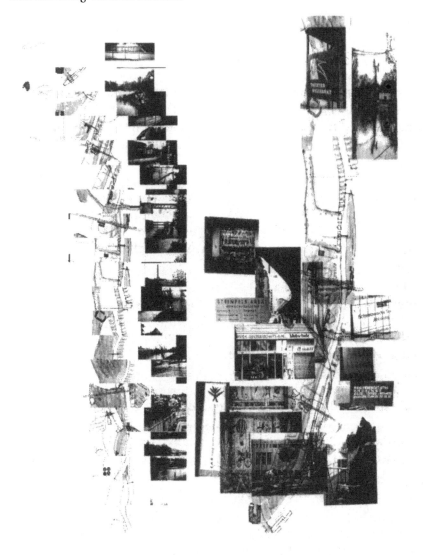

Figure 11.2 **The vision of a future river landscape, presented in the Stadtforum as a benchmark for the up-grading effort**

them. Its recommendations were submitted in the summer of 1997. A major outcome was that up-grading and development projects should be differentiated according to the specific potential of each part of the area. Furthermore they should be tackled in a so-called 'co-operative process'. Such a process would involve those with an interest in and a concern about the key issues in each area. The organising capacity should be coupled with a vision and strategy of sustainable urban development. Yet to formulate a vision is one thing; to implement it is another. Both activities may require a specific allocation of human capacities and organisational set-up (see, e.g., van den Berg et al., 1996, p.12).

The *Stadtforum* has led to the creation of two bodies. Their mission is to turn the suggestions made by the participants into concrete projects and action. The new task forces are the following:

- the *Impulse Group Zurich West* (IGA). It was intended to initiate all kinds of smaller projects to make the area more attractive;
- the *landowners group*. It engaged in a co-operative planning and development effort with the municipality of Zurich.

The initiative of the landowners which led to the Zurich West Development Project was characterised by some specific process-oriented *action lines*. These six lines are summarised as follows:

- to launch a *process*, by which various human capacities, private initiatives, state levels and departments of local government are brought into an optimal interaction for defining development goals, consolidating a short- and medium-term action plan and quickly achieving some results;
- to consider both real estate *market developments* and the *capacities of the state and the municipality* to finance infrastructures;
- to launch an *urban design*, which could mobilise interest from investors and individuals in the area and support the emergence of a new identity;
- to launch *area quality management*, which involves the municipality, the inhabitants and private enterprise in an interactive way;
- to launch *business site marketing* to promote a distinct structural policy for economic activities in the area;
- to emphasise the *attractiveness of the river landscape* and the 'cultural mile', which link the development area with the city.

The overall target was thus to mix a broad range of urban activities in an attractive cocktail of perspectives and land use. To achieve these goals, the landowners have addressed the departments of the municipal administration which were engaged in urban development. The pertinent co-operative process started in the spring of 1998.

Method: Observation and Action Research

The present study of the Zurich West development process is primarily based on qualitative analytical techniques, involving 'action research'. Action research means that the persons conducting the research are actively engaged in the development effort. This enables them to look at the process from the inside. In Zurich West the two authors had the opportunity to make firsthand observations of a lengthy and sometimes troublesome negotiation process. P. Güller was a 'complete participant' (process manager and mediator), whereas W. Schenkel acted as an observer, reflector and independent supervisor of the process manager.

However, a certain degree of methodological pluralism was applied to organise the storyline to enhance its relevance for network and output variables. The fact that participant observation involves a relatively intimate approach obviously implies that the researchers' presence may have an impact on the observed reality. On the other hand, evaluators are not just doing fieldwork out of a personal interest. They conduct their study on behalf of some decision-makers and to facilitate the work of other users of information. They can provide network participants with useful inputs. Creative fieldwork means that the evaluator is present on-site observing – talking with people, and going through programme records. No other methodology could have given the two authors such penetrating insight into the cognitions and emotions of negotiating city developers. At all times, they were able to describe the process – from unofficial draft papers to official final documents and public statements (similar approaches have been described by Waddington, 1994; LeCompte and Preissle, 1993).

The research procedure involved three steps. The first was to create the so-called storyline of the urban planning project by studying media products, literature, planning documents and legislation. The next consisted of analysing networks in quantitative and qualitative terms, including categories of belief systems, number and characteristics of actors, quality of relationships as well as network density and structures. The third entailed deriving network management guidelines and making organisational proposals. The analysis of the project aimed to increase insight into factors of success and failure of organising capacity.

Historical and Physical Context: Degradation and Regeneration

The problems of Zurich West have deep roots. Industrialisation in Switzerland was long kept away from the cities by the guilds. When their power was broken, some 200 years ago, a new wave of industrialisation – the production of machine tools, the first wave having been textile manufacturing – involved urban areas. Energy was still drawn from waterpower. The Zurich West site was originally at the fringe of the town. Later, when electricity made the factories independent of watermills, and in order to allow for spatial extension, the industries moved downwards in the Limmat River basin, to the then still rural periphery. The new location had a great advantage. It was directly accessible from the railway system, which was that time newly established. The area between the industrial site and the core of Zurich became one of the first workers' districts of Zurich as well as the centre of vocational education. Gradually, the city grew all around the area, making it the inner-city district it is today.

In the 1930s, the economic crisis hit the area badly. Since the 1970s, the de-industrialisation process has hollowed out what was left. It was turned into Zurich's backyard. In the early 1990s, the parts of the area close to the central station where taken over by the drug scene; that section became widely known as 'needle park'. In the mid-1990s, the drug scene was dissolved by the police presence and the introduction of health services. Soon regeneration efforts for the public space got started and a marketing commission for the area was installed. Between Zurich Main Station and the industrial site, the so-called cultural mile is in the making and co-operative housing schemes are being renewed. Obsolete industrial areas offer many opportunities for conversion. The previously large industrial complexes had once had their own transport systems. These estates have been broken up and now provide opportunities for new uses and urban life.

> *The area along the Limmat River, between the central railway station and the Zurich West Development site, is increasingly attractive for cultural activities. The start was made a long time ago, when the vocational schools where concentrated there, among them the school for design, arts and crafts. Recently dance schools and other clubs organising artistic courses have opened, as well as a series of progressive art galleries and studios and some trendy new restaurants. In addition, a large cinema complex and experimental stages of two of Zurich's main theatres were inserted into the old factories.*

By now, Zurich West has the potential of becoming a new, unique, modern, and attractive place for working, education, innovation, culture, leisure, and sports. The structural changes in Zurich's industry have overcome the former adverse impacts of manufacturing on its urban surroundings (Figure 11.3).

Figure 11.3   The envisaged cocktail of urban functions in Zurich West

High-tech industries are clean and thus compatible with almost any other use. The Swiss economy can only remain competitive if it concentrates on specialisation. Yet the recession of the early 1990s and the accelerated enhanced structural transformation processes in industrial production led to an unparalleled change in the Swiss real estate market for commercially exploitable land. After decades of scarcity of floor space for tertiary functions, the supply of rentable space started to exceed the demand by far. Rental returns and property values dropped considerably. In that light, it is not advisable to provide the city with more low-amenity office space. Rather, it is better to concentrate on conversion and the creation of highly qualified multiple-use premises.

Legal Framework: Deadlock and the Way out

A new policy-making approach was initiated by the city planners and the new political leadership. The new approach has its origins in a deep and unsolved conflict that arose in the early 1990s, about attempts to revise land-use regulations. The conflict was many-sided: left versus right, environmentalists versus the business community, municipal authorities versus cantonal government, the private sector versus the inhabitants etc. The conflict must be seen in a broader context. The world in which urban policy-makers find themselves appears to be dominated by uncertainty in many respects. Many of the effects of irreversible development disappear behind the limited time horizons of the agents concerned. The following problem dimensions have changed most dramatically in Swiss urban areas (Schenkel and Serdült, 1999):

- urban *problem structures* no longer correspond with federal, cantonal and local decision-making structures. The problem and impact structures diverge; i.e. the origin of a problem is not in the city but, for instance, in its agglomeration. Due to the complexity of the characteristically Swiss interlocking policies, cities are no longer able to solve urban problems by themselves. For example, cities are suffering from heavy traffic congestion, but appropriate solutions fall under the competence of cantonal and federal authorities;

- *top-down steering* by the state becomes more and more inefficient. City governments are overloaded, under financial pressure, and lack legal authority. They have no alternative than to rely on participation of new actor groups. And economic groups claim that urban policy will remain in a deadlock as long as the state is in sole charge of urban policy;
- *sectoral strategies* to solve policy problems become more and more inefficient. The origin of a problem is often located somewhere else, in another policy field. Traffic congestion, for instance, should not be tackled by building new roads. The solution lies in better land-use policy and fiscal incentives. New kinds of programmes have to be considered, namely cross-cutting, process-oriented, and co-ordinating programmes.

In 1991, the voters of the city of Zurich supported the municipal government's proposal for the legally binding zoning plan (BZO of 1991). But its implementation was blocked by a large number of legal claims brought by landowners and the political opposition on the bourgeois side. One of the main points of contention was the emphasis placed on the preservation of land dedicated to industrial use. The municipal government has applied such restrictive regulation to be able to prevent speculative development of office space. New use of industrial land should only be possible on the basis of a special agreement between landowners and the local authorities, the product of which is the so-called Gestaltungsplan (area master plan). The idea behind the industrial zones was also to prevent small and medium-sized industrial and business enterprises from being forced out by rising land values.

In order to overcome the blockade by legal claims and opposition regarding the new zoning plan, in 1995 the Canton of Zurich issued another zoning regulation (BZO Hoffmann). The new plan brought a much enlarged freedom of land use to wide areas of the city by means of the so-called 'centre zones' (*Zentrumszonen*). In the long term, these centre zones should be able to strengthen and secure the competitive position of the city and the canton among other European regions.

With regard to these centre zones, which allowed for high density and multifunctional land use, a heavy conflict arose between the municipality and the cantonal government. The conflict lasted for many years, involving 'old' ideological left-right and city-country argumentation. A parliamentary urban development commission gave evidence that communication among moderate politicians was still possible. Nonetheless collaboration between city and cantonal governments remained blocked.

However, over the last couple of years, there has been an opening in regulations and minds. Certain zones are allowed to have other uses, and the municipal authorities have realised that co-operation can break the deadlock. Both municipal authorities and landowners have an interest in revitalising the industrial districts and including them actively in the urban economic space. Successful co-operation between public authorities and private developers to realise the industrial district became one of the main goals of the municipality. The current political situation is favourable for large-scale conversion. On the one hand, the city of Zurich seeks to attract more people to live in the city. On the other hand, industrial districts are one of the most interesting locations of the city where large-scale business and sports or entertainment projects can be realised.

A first successful project applying a more co-operative planning approach took place in another old industrial area in Zurich, later named Centre Zurich Nord (ZZN). The initiative for restructuring, up-grading and developing the Zurich Nord area in Oerlikon was taken at the end of the 1980s by the landowners - Asea Brown Boveri (machine production) and Oerlikon Bührle Holding (machine tool production). A large part of the buildings and open areas used by the two firms had to be opened to new use, as lower (less sophisticated) layers of the production had to be abandoned and shifted to low-wage countries, rationalisation of production processes went ahead, new job requirements and workplace conditions emerged and industry got more and more tertiarised. This change became visible in the form of empty or dysfunctional production halls. In that state there was no chance to sell the land or the floor space. No new investor or customer would have looked for land which was allocated to industrial use only. Investors would have been afraid of polluted ground and the cost they would incur to restore it and break down the existing plant. And investors and customers alike would not have been attracted by the image of Oerlikon as a terra incognita, an unknown area with a bad image.

The land-owning enterprises initially made up their minds to maintain the Zurich Nord site. But they decided to reduce the amount of production. In 1988, to prepare for the re-conversion of the area, they created a common planning body under the name of Chance Oerlikon 2011. Already during the preliminary contacts between this private planning body and the municipality of Zurich, the essentials of the future development strategy were discussed. Is it true that the tertiary sector provides the highest return on investment in the Zurich region? Can high
Continued on next page

densities guarantee the necessary attractiveness and competitiveness of the area? What are the requirements in terms of infrastructure for creating a new vital urban area? How much public space does the area need, given the goal of a new multifunctional development? Such a multifunctional development was agreed upon in 1991, when the municipal government and the landowners signed a structure plan for Zurich Nord which set the main land-use parameters.

On the basis of the structural plan, a competition for ideas on urbanism was launched in 1991. In 1992, the four top-ranking consultants had to revise their drafts. There was a learning process on the side of the landowners and the representatives of the municipal government. They understood that the design which provided the most spectacular urban appearance by means of a large central park would not allow for stepwise development. Furthermore, such a large new 'green' nucleus would have required substantial changes in land ownership; the project could thus have run the risk of being delayed due to quarrelling landowners. It was found to be better if a series of single development units were identified which could be realised in a staggered way, individually equipped (stepwise return on capital investment) and newly used in accordance with the plans of the investors. It was also necessary to link the existing urban fabric with the new one, in order to follow the demand on the real estate market.

The balance of interests between the landowners and the municipality was reached by means of an exchange of land against performance. Land for roads and squares or parks was handed over to the municipality in exchange for a new land-use regulation which allowed for a broader range of land uses (previous zoning had strictly prescribed industrial functions) and higher land-use ratios. Pertinent agreements were laid down during a challenging process of negotiation. These arrangements took both a general form and the form of bi-lateral contracts between the municipality and each single landowner. Still, a minority of the landowners has been able to delay the process for some while by refusing to sign the general agreement. After signing all necessary agreements, the landowners returned to their own daily business and again became sharp competitors on the real estate market. Recent developments in Zurich Nord show that the market has reacted in satisfactory way to the newly created attractiveness of the site.

Development of Actor Networks

Concerning Zurich West, the *Stadtforum* took the initiative for a re-orientation towards more co-operation and voluntary agreement. The roughly 50 active participants of the *Stadtforum* have outlined the new directions in which complex urban planning processes should be steered. The process was neither time-consuming nor expensive (10 public meetings; 450,000 Euro). The *Stadtforum* succeeded in attaining its very basic objectives: consensus was reached on planning and communication guidelines, and the discussions were no longer shaped by political dissent. The

Stadtforum also had some shortcomings: too much formality instead of a workshop atmosphere; insufficient representation of specific groups, such as foreign inhabitants; lack of in-depth work, conflict handling and quality control; no link to concrete projects; too much emphasis on political de-radicalisation. The main thing was that 'the ice was broken' (Stadtforum Zürich, 1997).

Follow-up networks could profit from a broadly accepted and now established 'planning philosophy'. Uncertainty caused by legislation has not yet been resolved. But the newly formed actor networks are now seen as a chance to participate and to exert influence; a tradition of conflictive behaviour has been replaced by a movement toward 'creative trust'.

The Impulse Group

The first spin-off of the forum, the 'Impulse Group for Up-grading Zurich West' (IGA), consists of a network of local district representatives, landowners and municipal administrators. The IGA was established to initiate and evaluate projects that would make Zurich West more attractive. A means for the optimisation of the projects was seen in getting people to participate in the assessment and design process.

One of the first projects realised by the Impulse Group for Up-grading Zurich West was the opening of an old railway bridge over the Limmat River to the public. This new link forms part of a general improvement scheme for a recreational network in the inner parts of the city.

A second project is currently under way. It involves the up-grading of space under a long highway bridge which runs over Zurich West. This shadowy stretch has thus far been used for parking. It lacks an appeal as public space for adjacent housing areas. It could be used for occasional markets. It should become a zone of access to the previous industrial area which, is now being up-graded in its totality.

A third group of projects fall under the heading of 'Implantate' (which means insertion of specific value-adding activities or objects). The mayor has recently taken a substantial step in this direction by removing the statues of famous citizens of Zurich's history from the elitarian core part of the town and placing them in underprivileged areas in Zurich West for several months.

The follow-up process managed by the Impulse Group Zurich West has a short- and mid-term focus on stimulating pilot projects in the field of culture and public space improvement. In the longer term, it aims at a general urban up-grading, public-private partnership, and voluntary agreements.

The Landowner/Municipality Interaction

The second spin-off of the *Stadtforum*, the grouping of those landowners in Zurich West who dispose of real estate apt to conversion, was formed in the winter of 1997/98. Strategic networks can assume many forms. The networks can be arranged in three classes: public (sub-)networks, public-private (sub-)networks, and private (sub-)networks. Interaction between (sub-)networks is an important feature of organising capacity.

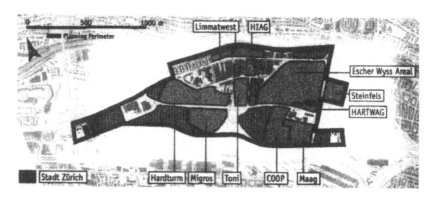

Figure 11.4   Planning perimeter and landowners

In Zurich West, the public sector was neither dominant nor passive; but it was eager that public and private actors complement and reinforce each other. In the course of the process the prevailing scepticism could be diminished. Then there was more and more willingness to co-operate and to agree on joint framework conditions. However, the degree of political and societal support for the process and its results remained unclear. Even though the actors were able to develop a joint strategy, they could not agree on an overall vision; federal and cantonal approaches to urban development were still lacking. The first phase of the Zurich West Development process was mainly financed by the landowners with the largest development potential (Figure 11.4). A second phase will focus on co-ordinated planning and realisation in each private and public area. The most ambitious plans concern a new soccer stadium and related commercial facilities. This project probably requires an intense network-building and managing process, taking project stakeholders, political groups, and nearby inhabitants into account.

Main Actors

Sulzer-Escher-Wyss Escher Wyss was founded in 1805 (Table 11.1). It is the oldest Swiss machine factory. Its goal was to industrialise textile manufacturing in Switzerland. After the merger to Sulzer-Escher-Wyss, SEW became one of the world's leading producers in several segments of the machine industry. Its activities include research and development, engineering, design, production, execution, commissioning, consulting, schooling and after-sales services. As the enterprise – due to economic modernisation – does not use the entire area in Zurich West anymore, today SEW also involves real estate management. That division is concerned with the development of the former Escher-Wyss area in Zurich.

SEW plays a certain pace-setting role in Zurich West. The municipality of Zurich and the company were able to integrate their goals in a site development plan. Such a Gestaltungsplan is based on a specific land-use agreement between the municipality and the landowner. The plan fixes the position, size and use of buildings. It was in the landowners' interest to add value to the site by a change of zoning, but at the same time they sought to preserve core industrial production.

SEW was taking part as a key actor in the development process of Zurich West. Its main interest was to couple its own projects with the up-grading process in the whole area. The main resources of SEW were its knowledge as a trend-setter and its financial backing It has limited interest in adjusting its own plans and projects in accordance with the up-grading process.

Technopark The Technopark of Zurich was opened in 1993. It serves as a melting pot and breeder for enterprises by allowing knowledge

Table 11.1 Actor characteristics

Actors representing	Functions	Resources	Network position
SEW	Landowner, producer, investor, trend-setter, area of cultural interest	++ knowledge, stock corporation in the background	++ core actor, mid-level representative, task force
Technopark	Landowner, investor, trend-setter	+ knowledge, story of success, stock corporation in the background	– peripheral actor, dependent on neighbouring projects, high-level representative
Steinfels	Landowner, investor, trend-setter, area of cultural interest	– investor corporation in the background	– peripheral actor, mid-level representative
Hardturm	Landowner, investor, stadium of public interest	++ added-value potential, land exchange, knowledge	++ central actor, core project, leading economic actor, high-level representative, task force
Maag	Landowner, investor, area nearby station of public interest	– hit by economic crisis, from production to real estate company, knowledge	++ central actor, core project, high-level representative, task force
Migros	Landowner, producer	+ rich company, interested in status quo, economic factor for city	+ peripheral actor, no concrete projects, low-level representative
Coop	Landowner, producer	+ rich company, interested in status quo, economic factor for city	+ peripheral actor, no concrete projects, low-level representative, task force
Toni	Landowner, producer	– unstable economic situation	– peripheral actor, mid-level representative

Actors representing	Functions	Resources		Network position	
Office for Urbanism	Planning and co-ordination, partly mediator, process initiator	++	represents government, knowledge, own research	++	core actor, high-level representative, link to city planning, task force
Urban Development Agency	Public information, social interests, interests of neighbouring areas	+	close to the mayor, knowledge, own research	++	high-level representative, link to neighbouring area, socio-economic inputs, task force
Public Real Estate	Landowner, investor, public interests	-	restricting double role, knowledge	-	peripheral actor, low-level representative
Public Transport	Transport supplier, public interest, implementation	+	knowledge, transport monopoly, existing plans, area development dependent on traffic regime	+	mid-level representative, technical and political input
Urban Infra-structure and Trans-portation	Planning, public interest	+	knowledge, development dependent on traffic regime	+	mid-level representative, link to superior policy level, technical and political input, task force
Environment	Environmental and public interest, examination	-	knowledge	-	high-level representative, task force
Green Area and Open Space	Environmental and public interest, planning, implementation	+	knowledge, own research, development dependent on free space	+	high-level representative, task force
Railway Company	Landowner, public interest, investor	+	national railway company, development dependent on station and railway network	-	peripheral actor, low-level representative, hardly present
Mediator	Planner, independent consultant, initiator	-	knowledge, no decision making	++	core actor, task force, process structuring

Source: Güller and Güller, 1999

and experience of science and economy to interlock. It allows for a new culture of co-operation between innovative enterprises through their spatial proximity. To create that synergy it is important to attract more science- and innovation-oriented institutions to settle in Zurich West. From the outset Technopark sought the co-operation of the surrounding landowners in order to strengthen its own competitive position. It also has established close ties with the Swiss Federal Institute of Technology and private investors.

Technopark Zurich is clearly a trend-setter in the area. It is promoting better accessibility to the area, better integration, and better working conditions. Its main resource is knowledge.

Steinfels After cutting back production and moving it to another location, the area of the old Steinfels soap factory area was converted into a site of residential, business, service, and cultural uses (e.g. Multiplex Cinema). The layout and the combination of uses contribute to the articulation of the urban structure in that part of Zurich. Like for Sulzer-Escher-Wyss, the development concept was fixed in a Gestaltungsplan (1988). Some sectors were based on the existing plant, while others will add new buildings to the area. The up-grading activities in the Steinfels area can be seen as an impetus for the whole of Zurich West.

In the most recent stage of the Zurich West development planning, Steinfels was co-operative but did not belong to the core group of Zurich West landowners, as its own conversion process was already well under way.

Hardturm Hardturm Real Estate Company is currently developing the former area of the Schoeller textile factory (where a first section of the housing and commercial compound has been completed) and the area around the Hardturm soccer stadium of the Grasshoppers Club. The development aims at capitalising on the attractive proximity of the river landscape for industry, commerce and housing. As for the site of the soccer stadium and training fields, the idea of the landowners is to create a sports and leisure area, including a new large-scale stadium for Zurich. Hardturm has a large potential to develop. Together with Maag, it held the entrepreneurial leadership in the process.

Maag Maag was founded in 1913 as a gear-wheel factory; it was growing steadily until the early eighties. At that time, Maag Zahnräder AG was severely hampered by a declining demand in its traditional market segments. The production of tool machines had to be given up in order to concentrate on the other segments (gears, pumps, real estate). In 1995, one segment of the Maag enterprise suffered great financial losses, which led to the sale of several core businesses, such as Maag gear and Maag pump systems. These high-tech companies are still located in the Maag area.

Maag Swiss Real Estate Group manages the development of the former Maag area. It now heads a project group whose aim is to turn the area and its surroundings into a neighbourhood of varied uses. The excellent access to public transport and highways makes the site a prime location for people-intensive land uses. Changes on the adjacent industrial areas are going to have an important and welcome impact on the Maag area itself and on the whole Zurich West area. The step-by-step transition to new uses is the basis of the feasibility concept.

Maag has a large potential to develop its area. It actively took part in the whole planning process of Zurich West. The main interest of the firm was to find investors and to bridge its own area development with public space projects (railway station, gateway to Zurich West). Its attitude towards the process can be described as sceptical but co-operative.

Migros, Coop and Toni Migros, Coop and Toni AG are widely engaged in food production, food (and non-food) distribution, and the administration of their core businesses. They intend to further concentrate these activities in Zurich West. In particular, their options for the future development of the area include specialised markets, leisure and fitness establishments, and restaurants. Their logistics needs will increase, which will have an impact on the volume of truck traffic. The three logistics giants will accept upgrading activities as long as their transportation needs are still met.

Migros and Coop were represented in the whole negotiation process. However, they took a wait-and-see position. The Toni Milk company was only partly represented in the negotiation process; in the longer term, its economic future in the area is uncertain.

Municipal administration The municipal administration was involved in the Zurich West Development Project, with several of its departments playing a key role. Those departments which engage in development and planning (the Mayor's Department and the Building Department) were deeply involved, as were the Finance Department with its Public Real Estate Section.

The departments involved in development and planning were key actors in Zurich West. They defended the public interest in the area's development and contributed to the structuring of the decision-making process. Substantial inputs were provided by the Agency for Urban Development, the Office for Urbanism, the Transport Section of the Infrastructure Department, and the Office for Urban Green and Public Space. The function of the Office for Urbanism has shifted from an operational level to a strategic level. Management stood above the individual interests. Specialised administrations (traffic, environment, green areas) were clearly promoting their own ideologies and represented specific interest groups (inhabitants, cyclists, ecologists etc.). The definition of area identity and planning concepts was relatively easy to achieve; administrations were open-minded and followed the landowners' ideas. As soon as it came to concrete figures such as the intensity of land use and the share of open space, it became a bargaining rather than joint solution-seeking process.

On the other hand, the Public Real Estate Section of the Municipal Finance Department represents the municipality in the capacity of a landowner. In Zurich West, a relatively limited – yet well positioned – part of the land is owned by the municipality. In general, one can say that public land ownership relates to specific public interests, such as providing public open space or low-cost housing. In addition, however, the municipal government is getting involved in the real estate market. Part of the city's real estate should be put on the market and treated as an asset in the municipal accounts. The municipal government is no longer willing to consider public involvement in area development as a mere provider of public functions.

In an overall view, the positions taken by representatives of the municipal administration were not far away from each other. Some of them did not yet see any reason to intervene; they will drop into the process when it comes to implementation. But in general, city representatives had a very positive attitude towards the process. A new administrative planning and decision-making structure can be seen as a particular side effect of the Zurich West negotiations.

Planners and experts The role of the planners and experts was in line with the new understanding of 'land development or change agents' (Kaiser et al., 1995). Land-use planning and decision-making can be seen as a big-stakes game of serious multiparty competition over an area's future land use. Planners can act as promoters, drafting the rules of the game and advocating community co-operation to achieve multiparty benefits. In the Zurich West development process they were expected to keep track of all stakeholders' interests, actions, and alliances. Due to the planners' credibility as experts, they had authority and opportunities to facilitate co-operation among competing interests (market, politics, society) in building a 'better community'.

In Zurich West, specific experts and three planning teams were invited to contribute to the vision-seeking process (Figure 11.5). The Dutch experts, for instance, acted as sparring partners for a) the municipality (city planners from Rotterdam) and b) the landowners (an investor group). The planning teams were based in Zurich/Berlin, Basel and Rotterdam.

The manager/planner of the whole process played a strong role in guiding the discussion and in motivating participants. His power to influence planning activities as such was, however, limited. The vision to look beyond immediate concerns and issues to see the needs of the future is a key attribute of the planner; visions and area identity were at least central topic in the negotiations. His main resources were knowledge and a deep understanding of the potential for the development of Zurich West; the financial basis was rather small.

Third parties such as smaller landowners, political bodies, and residents were not involved in the preparatory meetings and workshops. The intention was to let them join the process later on an

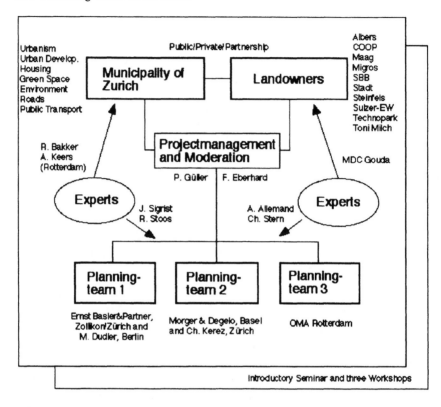

Figure 11.5   Zurich West Development Organisation

individual basis. Core actors such as Maag and Hardturm would contact them as far as they are affected by a concrete project. Municipal authorities are charged with information and promotion tasks. Inhabitants nearby the planned stadium started to formulate their needs and express their fears.

Much power was exercised by the core landowners (Hardturm, SEW, Maag) and the Municipal Office for Urbanism. Strong positions were also held by the Urban Development Agency and the Transportation Agency; they have supported policy change towards sustainable principles. However, in the early stages of the process, the environmental agency only had a corrective function. The achievement of the targets depends increasingly on policy programmes in the fields of private and public transport. Conflicts between economic growth and sustainability remain somewhat unresolved; it is not clear if these conflicts run deep or not.

Both the administration and the landowners stress that the tasks performed by the working groups were fruitful and had been carried out in a good atmosphere. Compromises have been made on a practical level, whereas quantitative targets and the vision for the whole area call for more negotiation and individual agreements. There may be some danger that the 'corporate identity' being built up throughout the first negotiation phase could weaken again in subsequent phases.

Structuring the Process

Zurich West is supposed to become an attractive inner-city area, designed and developed according to high standards of quality. The plans themselves might open new perspectives for both the area and the city. Making that vision come true will require considerable effort and financial resources, but especially time and flexibility. Because the growth will be spread over 10 or 15 years, it is subject to cyclical fluctuations. The plan must be able to absorb these ups and downs. Therefore, the planning process should be coupled with a wide range of special-interest interventions: an operational communication plan; a continuous social-return approach to meet social and political demands; a specific policy on public events; establishment of pilot projects and support for pioneers; use of art and culture as a motor for of new developments; and marketing and drives to attract business.

Few things are as difficult as coming up with an idea of how to revitalise Zurich West. In the first place, external factors are uncertain and thus difficult to predict. And a project like this one never proceeds exactly according to a previously outlined scenario. Nevertheless, in order to evoke a manageable image, the path to realisation has to be divided into different phases; certain elements may overlap with one another, and the amount of time necessary can vary. In the following section, the pioneering and negotiation activities are described in more detail. It is not yet known what the main contours, the final phase, or the completed city will look like. However, a well-structured and mediated negotiation process may raise hopes of success for the project.

From the beginning, it was clear that the management of change and complexity is best achieved when networks of organisations and individuals work together (Table 11.2). Local government provides an obvious framework for this corporate approach. That is because the local authority is already multifunctional. Moreover, it is increasingly geared to working alliances with other parts of the local public sector. In this context, we should stress that sustainable development is based on a core set of objectives for urban policy and presents the strongest possible justification for strong corporate policy. It is an outcome to which all public services and regulatory functions should be contributing. Furthermore, sustainable development calls for a business planning approach, specifying targets and measurable indicators.

The process was started with a joint inventory of the potentials and weaknesses of the urban site. The potentials include an integrative urban environment, excellent communications and infrastructure, an attractive image, access from the city by way of the Cultural Mile, the good state of the buildings, and the object of stimulating broad participatory effort. The weaknesses include the lack of attractive and contiguous open space, the relatively high land and rental prices, some areas with soil contamination, high noise levels, and so far little common marketing. The promoters of the process are the city administration, mediators, and parties with an interest in the industrial sites in transition.

The initial network was composed of ten real estate owners, seven departmental sections of the municipal administration, the project management and monitors, three planning teams, four project experts, and three planning experts from the Netherlands.

Later, relevant participants decided to establish a task force in order to work out a statement of intent. The task force consisted of three or four civil servants, four landowners and the project manager. Project teams and experts from outside were no longer involved in the process. Although the actors had modified their belief systems throughout the negotiation process, they remained suspicious of close collaboration with third parties and external societal groups. The task force decided to go public only after a basic statement of intent and planning guidelines had been work out. There was no interest in widening the range of knowledge and

Table 11.2 Negotiation process

Events	Topics
Preparatory meeting (29.5.98)	Information exchange, first ideas concerning co-operative planning process, clearing of positions, sceptical positions (waste of time, different planning stages), decisions concerning participants and project financing, sparring partners
Introduction workshop (29.7.98)	Existing projects, ideas, and interests; workshop organisation, tasks, contact with project teams
First workshop (18/26.8.98)	Formulation of goals and objectives, discussion of project proposals; first project team presentation, comments by experts
Second workshop (4.9.98)	Design of alternative plans; discussion on project team presentation, freedom of action, concepts, project contents, investments, and first implementation steps
Third workshop (21.9.98)	Design of alternative plans, final project discussion, decision on future negotiations, consensus and disagreement, composition of task force
First task force meeting (19.10.98)	Actor positions after workshops, land-use mix, short-term organisation.
Second task force meeting (20.11.98)	Public space, stadium, transport, preferred plan, discussions about statement of intent and development plan
Third task force meeting (7.1.99)	Detailed discussion on land-use mix, preparatory steps towards agreement
Fourth task force meeting (15.1.99)	Decision on agreement and development concept, implementation of adopted plan, first contacts in neighbouring areas, public information policy
Final workshop meeting (22.2.99)	Informing workshop participants and project teams about progress achieved by the task force, range of consensus
Final task force meeting (9.2.99)	Approval of the statement of intent, development concept and final progress report, involvement of political leadership
Municipal council information (24.3.99)	Information given to the municipal government by the responsible councillor, approval of going public, future organisation
Public information (7.4.99)	Joint press conference, results of the first phase, future steps, participation of third parties, separate negotiations and planning activities

Source: Güller and Güller, 1999

negotiation positions; note that hardly any people are living in the planning perimeter.

Planning Concepts

As mentioned above, the municipal administration and the landowners agreed on financing three project studies elaborated by three different teams of architects and planners (Table 11.3). At first, it was not clear if the teams were meant to compete for future contracts or to take part in problem-oriented discussions. The exchanges among the between teams, administration, experts, and the landowners was intense and sometimes conflictive. Each team had its own philosophy and thus came up with different planning ideas. No team was a clear favourite but all teams were active in the discussion, contributing to a better formulation of the area's identity and its planning targets. None of the projects they presented became part of the future task force negotiations. Indirectly, however, they helped make the ideas and procedures more concrete.

In addition, the municipal administration brought in three scientific analyses – concerning housing potential, zoning and related costs, and urban sustainability – that had been carried out by three independent consultants. These studies and their results were rather controversial.

At a subsequent stage, in the summer of 1999, Hardturm and the city authorities informed the public about four project studies concerning the new stadium (30,000 spectators) and its surrounding facilities (total 60,000 m^2). Four renowned developers (Brisbin Brook Beynon (Canada), Thyssen Real Estate (Germany), Skanska (Sweden), and Multi Development Corporation MDC (Netherlands), linked with well-known architects, presented rather different development plans and ideas. MDC was selected to do more work on detailed land use, the shape and location of buildings, accessibility, and economic efficiency. Future investments have been estimated at 200 to 250 million Euro. City authorities made it clear that no public financing can be expected, but land will be offered by the municipality and the Hardturm Real Estate Company. The selected developers will be responsible for the planning and financing. An architectural competition is going to be organised in

Table 11.3 Project characteristics

	Project 1 (Ernst Basler & Partner with Dudler)	Project 2 (Morger & Degelo with Kerez)	Project 3 (OMA Rotterdam)
Characteristics	Large-scale development project; new urban district with a large central park	Development based on existing trends and potentials; continuity	Large-scale new development based on substantial extension of transport infrastructures, based on functional relationships between the CBD, other areas of the town, and the agglomeration
Realisation	Interdependent realisation steps towards a well defined final appearance of the area; mid- and long-term perspectives	Step-by-step development; legal feasibility; short- and mid-term perspectives	Step-by-step conversion of the area; dependent on other city developments and the status of infrastructures; long-term perspectives
Infrastructure	Light rail, new station near stadium, improved city network	New station near stadium, streetcar, local network	New regional rail links and new high capacity intercity station, substantial new road links enhancing the position of Zurich West.
Land-use mix	Housing, working, services, culture, sports, high-tech production	Ditto	Ditto
Instruments	Strict masterplan covering the whole area, supervision	Flexibility, masterplans covering small-scale areas, adaptation of existing legislation	Multi-level network and activities, political and planning agreement covering the whole development process, quality management
Input to network discussion	Discussion on how to create identity	Discussion on what is historical context and existing potential	Discussion on Zurich's competitive position in a European context

Source: Güller and Güller, 1999

2000 and realisation is planned from 2003 onwards. Business and politics seem to have compatible interests, while potential opposition from social groups has actually not (yet) affected the planning process.

Policy-Making and Agreement

A Learning Process

The Zurich West network was formed by actors who were deeply involved in the decision-making process. Specific knowledge was not brought in from the outside. The learning potential was provided by identifiable forces which drove the policy process from one stage to another. These forces broke open the earlier legalistic top-down approach, paving the way for the use of communication instruments and engagement in a variety of interacting cycles. Learning was possible on the basis of a problem-oriented bargaining system. The participants were willing to exchange information and to take contradictory belief systems seriously. Learning was also supported by the aspiration of increasing social prestige, the intention to avoid expensive and time-consuming political disputes, and the expectation of an overall decrease in implementation costs.

Paths to Agreement

The process was more than bargaining and looking for compromise. It was also about changing perspectives and managing the context. Several policy fields and even new conflicts were taken into consideration. Hereby, the lack of clarity with regard to what sustainable development could mean had certain advantages. The vision of future land use was based on three pillars:

I   market values;
II   social values;
III ecological values.

Sustainable development implicitly served as the foundation on which these pillars stood. It has allowed groups with different and often conflicting beliefs to identify some common ground upon which concrete action could be designed.

Members of coalitions did not necessarily belong to the same institutional body. Some coalitions could be seen as sub-networks, held together by common belief systems. Co-operative principles partly based on sustainable urban policy guidelines have led to better network management among the relatively small number of actors. Participants have realised that gaining community support, understanding, and 'ownership' of plans is a necessary but not a sufficient condition of successful implementation. It must be followed by systematic programmes to ensure that adopted plans are used in operational decisions (Kaiser et al., 1995). The participants agreed that entrepreneurial planning is not just about recreating the market. Is has the wider goal of getting people and vitality back into neglected areas; it aims to inject them with life and excitement. Some actors have developed a collective responsibility, self-control, and joint initiatives. Reflexive steering was somehow the only instrument which offered long-term and durable solutions. Ecological sustainability was not a target as such. But the actors' acceptance of future uncertainties and planning flexibility is an inherent factor of sustainability, understood as a normative, process-oriented and long-term concept.

Uncertainties

Uncertainty about future developments on the real estate market were omnipresent during the Zurich West planning effort. It can be seen as a lack of clear options, which blocks the decision-making process. But it can also be seen as an important process variable. In this perspective, government and business actors were willing to adopt a wide application of the precautionary principle, arguing that the costs involved in taking action under conditions of uncertainty are high. The key question in analysing and developing political and economic policy options for the transition to a new area development is that of flexibility, where flexibility is defined in terms of uncommitted potential for change; and this was exactly the case in the Zurich West area. Many development perspectives and framework conditions were uncertain. This fact enabled the participants to create flexibility for future projects; options for change became a target of the negotiation process, even for the extension and shape of public space.

By-passing Upper State Levels and Legislation?

Looking at the entire process from the *Stadtforum* to the Zurich West private/public negotiations, it seems that unstable relationships with clearly conflicting elements and rather formal or hierarchical characteristics turned into stable relationships with elements of network management and network building. However, multi-level co-operation was still lacking. Some measures may remain ineffective if co-ordination with supra-ordinated levels is not enforced with the same resoluteness. Furthermore, there was some disagreement about the meaning of covenants. Are they compatible with legislation? Others stressed that the law is not and never was under discussion; it was just given a double function. The government can simply enforce the law or it can use a law as an argumentative source of power, while negotiating and concluding 'voluntary' agreements.

We should, however, not forget that covenants are not a fashionable policy instrument in Switzerland. There is no tradition of direct bargaining between small groups of city officials and real estate representatives. On the other hand, Swiss federalism offers many possibilities for consensual solutions, although direct-democratic elements set certain limits to consensus-seeking processes. Ultimately, it is up to the political or social majorities to decide, sometimes after highly polarising public debates and political campaigns. The Municipality of Zurich is ambitious in this context, but it needs more time to develop and redefine the voluntary approach.

Deregulation was never a goal in itself. It became part of a broader strategy to achieve a more positive relationship between economic and social developments. However, discussions on the application of the law will have to be revived at a later stage. The negotiation process was launched as a stopgap solution until new land-use regulation can be put into force. It can be seen as steering from a distance and creating co-responsibility. There is no legal power to enforce the commitments, but the participants agreed that covenants are based on a moral obligation which is not so easily overthrown. In the Zurich case, the agreement was not seen as an 'easy' alternative to avoid regulatory burdens; rather, it was greeted as a way out of the deadlock.

Statement of Intent and Development Concept

The first phase of the Zurich West development process led to a written document. It consisted of a project description, the mutually agreed upon statement of intent, and the development concept (Table 11.4), with relevant land-use and transport plans (Güller and Güller, 1999).

The statement of intent stands for a certain policy change and is the result of the learning process. In a nutshell, it says the following:

- *the objectives* are efficient co-operation with a view to long-term attractive development and urban quality. The market potential as well as the political and legal feasibility are to be considered. A proposal for the conversion of the statement of intent to a politically sanctioned directive should be submitted to parliament by mid-1999. By that time, the area-wide concept has to be complemented by property-related rules. Each landowner shall know about the legal backing of his development plans;
- *the guidelines* aim at adding value for inhabitants, landowners, and the business community. Sustainable development must be understood as an attractive and market-oriented land-use mix in terms of international competitiveness, environmentally friendly infrastructures and open space, as well as participation, transparency, security, urban quality, legal compliance, and resource management. Procedural principles demand flexibility, legal certainty, efficient decision-making, and specific forms of infrastructure financing. Basic principles have been uncoupled from the legal framework. The emphasis is on market conformity, co-operation, sustainable development, and information policy;
- the landowners and the representatives of the municipal government shall form a *'conference for co-operative development planning of Zurich West'*. Operative bodies shall elaborate financing and marketing models for the infrastructures.

Table 11.4 Development concept and learning process

Development concept		Policy change
Targets	Attractive new city district, flexibility and stepwise approach, land-use mix and identity, new image	Advantages of the area are exposed and underlined, disadvantages no longer mentioned; public space becomes part of the planning process
Basic planning principle	land-use quality and quantity have to be defined under criteria of quality needs and public–private partnership concerning infrastructure and open space supply	Joint approach to private and public interests; qualitative development
Quantitative area-wide land-use regulations	floor space index between 2.0 and 3.0; 20-30% housing; 5 m² and 8 m² open space per working place resp. inhabitant; no compensation concerning housing and open space ratios between existing and new land uses	Agreement on the basis of the entire area; no norms concerning existing buildings; open space and specific floor space indexes need further definition
Urban development	Zurich West identity, key projects, new public space, quality and urban density	High-rise buildings should become possible; key projects defined
Open space	green network, green parks, green link to neighbouring areas	Space accessible to the public; open space quality as a marketing factor
Housing	Substantial share of housing, qualitative concentration, appropriate infrastructure	Flexibility, new housing forms, housing for medium and higher incomes
Workingplaces	Attractive and high-quality work places; new identity	Consensus achieved; mix of small and medium-sized enterprises
Traffic	Public transport projects, link between road traffic capacity and availability of parking space; efficient infrastructure for logistic enterprises, high porosity for cycling and walking; link to river and culture mile	Private financing as an option; public transport promotion and private traffic reduction supported by all participants; consensus achieved; however, superior state levels not yet involved

Source: Güller, 1999

Authorities co-operate with landowners while developing legal frameworks. Third parties will be included step by step;

• the *development concept* is a first step towards an implementation plan. It links private projects (working, living, sports, leisure, culture, logistics) and public infrastructure planning (schools, green space, public space, transportation). It involves respect for interests and developments in the adjacent areas. There is a change from individual targets towards a common identity and interdependent targets. Sustainable development is 'positively' formulated in order to attract potential investors.

Public Open Space: a special Issue

Open and public space was a central issue. It was planned to extend from the Central Station along the river landscape to the Zurich West area, including the 'cultural mile'. It was also planned to connect private space, and to provide the new area with an identity by creating 'green' niches and networks. Some streets may become 'green boulevards'. Small open spaces and pathways will be established between buildings. Important up-grading efforts have to be done nearby the railway station Hardbrücke. In areas such as Hardturm, Schoeller, and SEW, an opening towards the Limmat River will be considered.

Public space should thus be looked at as one of the key subjects of negotiation between various actors involved in the area development process. The following dimensions of public space may be taken into account:

• public space as a means for closing up an area which, during its industrial use, was fenced off, making it a kind of *forbidden city*;

• the *accessibility* of public space to various segments of the public (partial openness, total openness). Public space in Zurich West can, for instance, be open to all who work there, but it would have no specific function for the urban population at large. Alternatively, public space in part of Zurich West could be part of a commercial establishment and function as a 'mall', involving closeness during the night and against undesired visitors such beggars, drunkards, etc. Another possibility is that public space would be open to all strata of the population and all kinds of users;

- public space as the *expression of a public will* for providing functions of public interest (schools, parks, functional links between areas, pedestrian areas, etc.);
- public space as a *backbone of urban design*, as a quasi-stable element of a development scheme, in contrast to other land uses which respond to market flows. Alternatively, public space may be so closely linked with the private urban functions it serves that its design would depend on private investments and developments;
- public space as means *to balance out the interests* of various groups which are actively involved in or touched by an area development project (barter subject);
- public space as a *subject of zoning regulations* or of a covenant among public and private partners in development, which involves financing modes;
- creation and maintenance of public space as a public, private or shared *task*.

Evaluation

An evaluation of the Zurich West Development process from the outside fails to answer the question whether the voluntary approach has actually contributed to a change of belief systems and to real sustainable thinking. But the statement of intent made by the landowners and city representatives has reduced uncertainty about the future course of government policy. There was hardly any actor who really wanted to give up the agreement approach. Yet some demanded better procedures and firm political guarantees. The statement of intent provides a good basis for long-term policy. Nonetheless, it has some shortcomings concerning openness for political and societal groups and democratic control.

The landowners admitted that the targets set in the agreement seem rather vague. But to deal with this is part of a learning process in a broader context and should be judged over the long term. Belief system changes could be observed, although the landowners were not willing to play a pioneering role.

Conclusions and Outlook

The *Stadtforum* for up-grading Zurich West has fulfilled the network conditions of variety but had deficits concerning stability. It has included all kinds of actor groups, was mediated and accompanied by experts, and has covered an extended planning perimeter. But its active lifespan was relatively short, and it had no authority to make substantial decisions. The *Stadtforum* was designed as an 'icebreaker' and opened the way to intensified collaboration on many dimensions of urban policy.

The initial phase of *the Zurich West Development Project* had a clear workshop atmosphere and links to concrete projects. However, its representiveness, legitimacy, and transparency were not beyond doubt. Disagreement concerning governmental and economic points of view had not been overcome. There is a fair chance that further negotiations will be successful as long as they remain oriented towards the common umbrella which the network participants created during the period of policy formulation. However, the local government will have to align its strategies with cantonal policy, as cantonal representatives were missing during the negotiation process. It is to be expected that new negotiations will be necessary in the implementation phase.

There is no easy answer to the question whether prescriptive and proscriptive legislation is more efficient than negotiation in networks aiming at self-imposed responsibility and voluntary agreements. If we look at the criteria for changing an urban policy regime, it looks like the Zurich West Development Project was switching between 'old' urban planning traditions and 'new' urban planning concepts.

- *what? Historical spatial and political structures* play an eminent role. Old and new identities had to be redefined and developed on the basis of historical structures. Urban quality as a general source of identity should be oriented towards sustainable accessibility, private business, good working conditions, and overlapping private and public space. Public transport and urban sustainability become a promotional factor. Conclusion: *Discussions on the identity and urban quality should be coupled with a stable and institutionalised discursive process towards "What is the city's future with respect to the quality of life and role within Europe?";*

- *where to? Objectives* and *goals* of urban areas and regions should be formulated in comprehensive strategies ('words') rather than in plans. At the beginning of the process, *ideas and visions* were focused on land property. During the negotiations, joint ideas and visions were formulated at an early stage by the landowners and the municipality. At the end of the first phase, participants were not yet willing to realise an overall vision. They were ready to interconnect their own ideas and to develop these ideas under a common umbrella. Plans served as a tool to formulate common definitions. Conclusion: *Urban quality and 'sustainable visions' should become the subject of long-term, institutionalised and broad discussion;*

- *how? Confidence* replaces regulation as a basis of planning. Confidence and a voluntary agreement had to break a legal deadlock. Instrumental steering was communicative and mediated, but conflicts about regulatory instruments remained unsolved. The call for area-wide, long-term and comprehensive regulation has been abandoned. Close-to-market steering was necessary to attract investors and to put into force comprehensive regulation taking into account both strict norms and flexible negotiation. Public-private partnerships was supposed to be a core factor in securing financing and long-term planning stability. Conclusion: *Communicative steering should become more stable and less interrupted. It cannot replace legal certainty and specific legal solutions (Gestaltungsplan);*

- *who?* There is a *new range of actors.* New organisations are created within or outside the political-administrative system (project groups, public-private partnerships, intermediate organisations, scientific experts). The range of actors was limited to public authorities, private enterprises, and scientific consultants. New societal groupings played no important role, due to the small number of inhabitants in Zurich West and to some degree as a result of the amount of private property. Opposition came from the inhabitants living nearby the new stadium both within and outside the planning perimeter. However, coalition building was clearly shaped by network management and discursive negotiations. Conclusions: *Urban quality, green areas and open space should become the subject of bargaining between project promoters and neighbourhoods. Uncommon coalitions, e.g. between landowners and residents, are assumed to have supportive functions;*

- *when?* Dealing with *time* and *deadlines* is becoming a more and more important motor of regional restructuring. There was pressure – largely coming from the private landowners – to define clear time horizons and deadlines. Conclusion: *Stepwise realisation must be supported by a clear time frame and deadlines;*
- *with what?* Besides financial resources and political support, successful spatial management projects need *promotional spirit* and *leadership*. This can come from the specific competencies of organisations or from the charisma of individuals who act as project-pullers. The promotional spirit and leadership could be revived by a relatively new city planning crew. Elite behaviour was sometimes competitive, sometimes co-operative, but consensual legitimacy became a vital norm in the planning process. Conclusions: *Elites should make better use of information coming from sources outside the network such as residential groups, future inhabitants, and examples from abroad. Social and political support should be seen as a central resource (knowledge) and systematically considered in the networking.*

Current processes dealing with urban development implicitly or explicitly consider new forms of policy-making. These new forms include co-operation instead of confrontation, joint target setting instead of top-down steering, and sustainable up-grading instead of purely economic recovery and modernisation. However, these efforts could not change the fact that developers expect land-use planning to facilitate their difficult and risky efforts to meet the market demand for developed land. They like real estate markets with minimum constraints, maximum flexibility and certainty of outcome, and vigorous public funding of infrastructure. Government tries to intervene in the development market on behalf of political and public interests. Public service must show regard for equity and due process for all citizens, as well as good standards, information, access, and redress for actual consumers (Hill, 1994).

Today, there is less conflict. But local government in Zurich has become a less dominant player in the local public sector. On the other hand, local government, together with its network partners, seems to be stronger in finding 'good overall solutions'. One can observe a clear shift from the use of power to more co-operative behaviour, a shift from non-management to network management and network structuring, and a shift from rationally oriented evaluation to learning-oriented evaluation. These developments were influenced by the prevailing criticism of 'over-regulation' in urban policy. However, there was not much willingness to build future-oriented networks. Swiss implementation processes are dominated by public authorities and law enforcement. In future, there should be more room for additional negotiation during the implementation process, permanent public information, and participation of the affected and interested groups.

In Zurich West, it is proposed to replace the task force by a conference for co-operative development planning. The task of the latter would be to prepare for future steps. It would pursue targets such as better legal certainty, interrelations among individual projects and among neighbouring areas, concepts of local and regional public transport, collaboration between other city development organisations, and common marketing.

References

Güller, P. and Güller, M. (1999), *Kooperative Entwicklungsplanung Zürich West*, Synthesebericht. Hochbaudepartement der Stadt Zürich.

Hill, D. M. (1994), *Citizens and Cities. Urban Policy in the 1990s*, Harvester Wheatsheaf, New York/London.

Kaiser, E., Godschalk, D. and Chapin, F. (1995), *Urban Land Use Planning*, University of Illinois Press, Chicago.

LeCompte, M. D. and Preissle, J. (1993), *Ethnography and Qualitative Design in Educational Research*, Academic Press, Inc., San Diego.

Schenkel, W. and Serdült, U. (1999), Bundesstaatliche Beziehungen, in U. Klöti (Ed.), *Handbuch der Schweizer Politik*, NZZ Verlag, Zürich, pp. 469-507.

Stadtforum Zürich, Schlussbericht, August 1997.

Van den Berg, L., Braun E. and van den Meer, J. (1996), *Organising Capacity of Metropolitan Cities*, European Institute for Comparative Urban Research, Rotterdam.

Waddington, D. (1994), Participant Observation, in C. Cassell and G. Symon (Eds.), *Qualitative Methods in Organisational Research*, Sage, London/ Thousand Oaks, pp. 107-122.

12 Akerselva Environmental Park: Urban Transformation by Chance or by Governance?[1]

KNUT HALVORSEN

Introduction

This book is concerned with the transformation processes taking place in European cities and the governance systems guiding those processes. This theme implies a focus on both the causes and the effects of the changes in the urban fabric. The changes are observed in the physical, industrial, demographic and social dimensions but also in the urban planning, management, and governance systems themselves. The challenge, and thus the main question raised in this chapter, may be formulated as follows: Is it possible to influence the development of our cities in the intended direction(s) given the new conditions of a market-driven and decentralised economy?

This contribution presents a 'social democratic' case, the Akerselva Environmental Park project. Here the term 'social democratic' means that the project was dominated by public ownership and a public planning process. But at the same time, the project was also one of the first clear cases of the shift away from the old 'social-democratic planning paradigm'. It was a process whereby certain elements were drawn from new liberal ways of organising urban projects. If a new planning model was evolving, the economy was evolving even faster. During these years, the Norwegian economy was susceptible to strong forces of change. At

267

that time (the 1980s), the transformation process in the urban fabric of Oslo was probably stronger than in most other regions in Norway. The objective of this chapter is to shed light on the project's strong social democratic foundation and on that basis to argue that strong public ownership was a prerequisite for the success of the Akerselva Environmental Park project.

However, this chapter also shows that the project evolved along the same lines as urban management, which tries to integrate socio-economic, environmental, institutional and financial aspects. Even though it was based on historic public investments going back to the 1920s, it is a case of a project taking an incremental approach to public decisions. It broke with the traditional social democratic model based on a large-scale urban master plan. Yet it differed from other European and North American (popular) trends by not being a bottom-up and public-private partnership process. In fact, it turned out to be a hierarchical public management project, run and controlled by a relatively small group of high-level bureaucrats. The Akerselva Environmental Park is, however, a clear case of network governance. In contrast with the case of Zurich West Development (discussed elsewhere in this book), the networks that evolved include various public agencies and levels of government, the links are not so clear between private and public actors.

Akerselva Environmental Park is also an example of how environmental urban policies and modern urban industrial development can be compatible. By improving infrastructure – that is, making traditional investments in the transportation area, culture and recreation, education, public services, public spaces, etc. – it is possible to create the kind of milieu that is in demand by the new growth industries.

Theory and Methodology

The project of Akerselva Environmental Park was carried out from 1987 to 1990. Thus, it is possible to evaluate both the short- and the long-term effects. At the same time, the project is recent enough for the actors to remember the process relatively well.

This particular study took place through most of 1998 and the first half of 1999. It was divided in two main parts. First, it consisted of an analysis of the transformation process of the past: How did the urban fabric of the Akerselva River Basin change both physically and functionally over time? Secondly, it consisted of an analysis of the planning process.

The theoretical approach falls under the governance/management perspective (see section 'The economic history behind the physical and functional transformation of Akerselva River Basin'). This means that the approach deals with governance questions related to the problems of the 'centre-less society' (Luhmann, 1982). How do the many agents and institutions interact when there is no clear agenda and when they have different goals? A decentralised economy calls for different systems of governance, which is a weaker form than government.

'Governance (…) is a way to directly influence of social processes (…)' (Rhodes, 1997). In this approach, governance is clearly different from public management. In Rhodes (1997, p.XI) governance refers to self-organising, inter-organisational networks with the following characteristics:

- interdependence between organisations. Governance is broader than government, covering non-state actors. Changing the boundaries of the state means the boundaries between public, private and voluntary sectors become shifting and opaque;
- continuing interactions between network members, caused by the need to exchange resources and negotiate shared purposes;
- game-like interactions, rooted in trust and regulated by rules of the game negotiated and agreed by network participants;
- no sovereign authority. Networks have a significant degree of autonomy from the state and are not accountable to it. They are self-organising. Although the state does not occupy a sovereign position, it can indirectly and imperfectly steer networks.

The state has lost much of its authority and position, because it is willingly giving power away in the shift towards market liberalism. In the last 10 to 15 years, we have seen a wave of decentralisation and outsourcing of many of the state's former roles and services. An increasing share of resource allocation is now taken care of by the market, which is nothing less than a larger number of private and semi-private (and even public) agents who are 'maximising their own utility or profit functions' without regarding the public interests – or, in this case, the city as a whole. Some describe this as 'the centre-less society' (Luhmann, 1982).

These are international trends; but how true are they for Norway and the city of Oslo? The study that underlies this chapter was financed by many of the same public agents who were key actors in the Akerselva Environmental Park project and some who are now becoming key actors in the same area. These include The Ministry of Environment, The Directorate of Public Construction and Property (Statsbygg), the Norwegian Research Council and the Ministry of Local and Regional Development. The City of Oslo participated by providing subsidised data (detailed map information) and manpower through out the project. The most important private real estate developer, Avantor, contributed likewise, providing information and taking part in meetings and seminars.

The methodological approach was designed to meet the needs of different parts of the study. The physical and functional transformation was studied by analysing planning maps and databanks describing buildings and infrastructure and changes through time in the defined area. This part of the study was done by the Oslo School of Architecture. The main data source was the Department of Planning and Building (Plan og bygningsetaten) of the City of Oslo. The analysis of the institutional process was structured as a qualitative study by the author of this chapter. The main sources of information were planning documents, reports, maps, governmental memos and, most significantly, interviews of the main actors involved in the process.

An important activity in this second part was to bring the main actors together in order to see how they interacted, but also to pinpoint the common 'truths' that they could agree upon (in a public setting). Therefore, after some preliminary studies, the projects started by inviting most of the actors to a seminar (in May 1998). At this seminar, they were asked to give 'their side of the story', by making short presentations on how they interpreted their role, the role of others, and key factors of success and possible failure. In the period after the seminar, some of the participants were interviewed again, while some of the actors who had been prevented from attending the seminar were interviewed.

Later – in October 1998 – the same people were invited to a second seminar. The purpose was to evaluate the preliminary results of the study that had been conducted in the interim. By this time the main findings, analyses and interpretations of the key actors had been worked out so that some preliminary conclusions could be presented. It was interesting to observe some of the reactions to these conclusions. It proved that, in cases like this, there are often many interpretations of reality. Some of the reactions suggest that most of the actors are players in ongoing games and existing networks between the same institutions.

The Transformation of Akerselva River Basin

In order to understand the setting and the circumstances, in which this particular urban development took place, one should know something about the external forces of change and the history of this particular urban area. This section gives a brief description of the background of the project. The Akerselva River Basin is the functional part of the River Akerselva. It stretches 10 km from the north of Oslo and Lake Maridalsvannet, running through the centre of Oslo and into the Oslo Fjord to the south (Figure 12.1).

As in many other European cities, the river represents one of the strongest geographical starting points for industrialisation at a national scale. Since the 15th century, the river had been one in the main factors of the industrial development of Oslo and, in fact, of Norway. The river was used as a source of power for the first mills and thus formed the basis for the first industrial revolution in Norway. The river had already been one of the major economic forces behind the development of the city of Oslo in the Middle Ages. The export of timber and the location close to the great forests of Eastern Norway gave impetus to the development of the city. Due to these resources, Oslo gradually became the most important city in Norway (surpassing the old Hanseatic merchant city of Bergen).

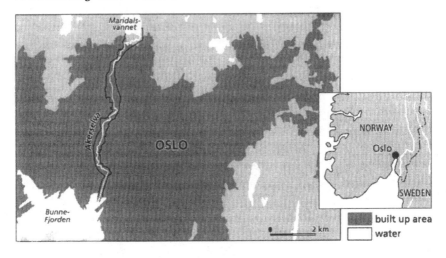

Figure 12.1   The river basin and its position in Oslo

During the first decades of the 20th century, Norway - and Oslo in particular - had experienced a spurt of industrial growth. Mechanical industries of all kinds expanded their production. Those were the days when Norwegian waterfalls were exploited at a large scale and the shipping industries were booming. Income increased and a large part of it was passed down to the fast-growing working class. This created new markets for the domestic and consumer-oriented industries. Products like textiles, food products and soap became the new growth industries in the Akerselva Basin. The existing factories from earlier periods, and the large sections of working-class housing areas on the east side of the Akerselva, were important location factors for these 'new' industries of the 1920s and 1930s.

During the fordist era of the 1940s and 1950s, the Akerselva River Basin expanded its economic position. In the 1960s, the industrial growth in Oslo simply outgrew the limits of the river basin. Industry looked for cheaper and more easily developable land in other parts of Oslo and in its neighbouring municipalities.

In the early 1980s Norway went through a fairly strong recession. The economic slump accelerated the decline in the old industrial structures of the Akerselva River area. Between 1980 and 1992, Oslo lost 20,000 jobs (38%) in manufacturing (Kann and

Halvorsen, 1995). As a consequence real estate became cheap; old buildings, empty factories, warehouses etc. could be bought or rented for a low price. The location was still very attractive, being close to the old parts of the city. During this period of 'creative destruction' (Schumpeter, 1943) the foundation was laid for the emergence of new industries. In the last twenty years of the century, there was strong evidence of the new IT and media industries rising from the 'ashes' of earlier industrial revolutions, leading to a mix of land uses. Today, several modern industries are growing within the old manufacturing industrial infrastructure. In the upper (northern) part, where the old steel mill Spigerverket used to be, the area has successfully been transformed to accommodate new 'clusters' of both IT-related and media firms. Most of the commercial film and television series in Norway – like soap operas, game shows and commercials – are now produced there. In the same area where once about 2000 jobs were lost, more than 5000 new jobs have been created in the new high-tech service industries.

Also along the lower reaches of the river, industrial regeneration is taking place, though much more slower. Even if the improvements have made the area more attractive, both for industry and the inhabitants, problems abound – particularly on the eastern side. That area still has some of the worst poverty, health and drug problems along with the largest proportions of immigrants from Third World countries found anywhere in Norway. The Akerselva River thus still marks a sharp difference in living conditions between the poor east and the wealthy west, a division seen in many other European cities.

One of the main results of the transition process of the 1980s and 1990s is however that most of the changes has taken place *inside* the physical structure – that is, inside the buildings and the old factories in the area. The most dramatic changes in the Akerselva River area are thus *functional* changes. This is mostly due to the influence and co-operation between the environmentalist – or rather the conservationist – interests in the city politics (at both city and state level) and private developers. There are also extensive physical changes – particularly in the large area controlled by the private real estate developer Avantor in the northern part of the river basin. That is where one will find the modern architecture – glass and new materials, which are similar all over the world – side by side with old factory buildings from the last half of the 19th century.

Planning History: from Government to Governance?

To explain some of the results of the park project, we have to go all the way back to the 1920s, when the first investments were made to create an environmental park. The strong emphasis on growth in the second half of the 19th century and the first decades of the 20th century led to severe environmental problems in and along the Akerselva River. This had become clearly evident by the early 1920s. Most planners and politicians – both on the left and the right – agreed that something had to be done. The river had become an open sewer as a result of the pollution from the homes of thousands of working families and the many factories located along the river. Akerselva had become a problem, both because of the environmental contamination and the health problems related to it.

To resolve that problem and to meet the growing need for effective mass transit, some parties came up with the idea of building a rail system running north and south through the city *on top* of the river. The proponents found that the idea would have covered the river (turning it into a real sewer) while solving some of the transportation problems prevailing at the time. The idea was presented in the early 1920s. It was the chief of city planning, Mr. Røhne, who prevented it from going ahead. Røhne saw the need for developing 'green lungs' especially for the thousands of working-class families living in the area. He led the politicians to believe that such a transportation project would be too expensive. He eventually convinced them of an alternative strategy; the fast-growing city needed green areas and parks, so the city should instead buy more land and clean up the riverbanks. Thus the idea of a turning it into a park (or giving it some park functions) stems from social democratic principles (although Røhne was not a politician) of the early 1920s.

The main strategy in the decades that followed was thus to buy property, little by little, year by year, along the river. This was the main social democratic strategy in Norway from the late 1930s to the early 1980s. Public ownership of the main infrastructure and semi-public ownership (co-operatives) in the housing sector were the strong pillars of urban policy. The maps in the figure 12.2 show how the authorities (mostly local) in the park area (excluding the buildings), that later became the Akerselva Environmental Park, increased its direct ownership from 169,800 m² in 1948, to 257,967 m² in 1962. By 1980 the holdings had increased to 363,018 m². Interestingly the purchase of property went on through most of the 1980s and the 1990s. In those decades the government was controlled by parties and coalitions of parties to the moderate right side of the political landscape – both in the city of Oslo and on a national scale. By 1998, the public share had increased to 434,340 m² or 85% of the park area.

The maps in figure 12.2 show the distribution of public versus Private Park area, not the amount of investment that went into the different parts over the years. During the 1950s and the 1960s, public money started to flow into projects to upgrade the areas that had been purchased in previous decades. The riverbanks were cleaned up, and small parks and green space were developed.

As a result of this upgrading people became more aware of the quality – or rather the potential quality – of the river basin. Environmentalism was not very strong in Norway at the time. These years were the heyday of the old-type fordist manufacturing. Industrial growth was the main aim, and this left little room for environmental considerations. Luckily for the Akerselva River Basin area, its industry lacked the space for further expansion. Many of the larger firms looked for more space and cheaper labour outside Oslo. This relieved some of the growth pressure on the area.

The main projects that concerned the city planners at the time were large infrastructure works. The city was growing fast and was in desperate need of housing and transportation systems - both mass transit, like the city metro, and large road systems. This was a period of extensive suburban growth, and the city was expanding outwards from the centre. One of the largest infrastructure projects was the highway that would have cut through the very heart of the lower part of the river basin. This project was mostly developed and 'owned' by planners but it was stopped by politicians before it got off the drawing table. It represented a large setback for large and 'old-time' social democratic infrastructure projects. The year after that – and into the 1970s – it became more and more difficult for

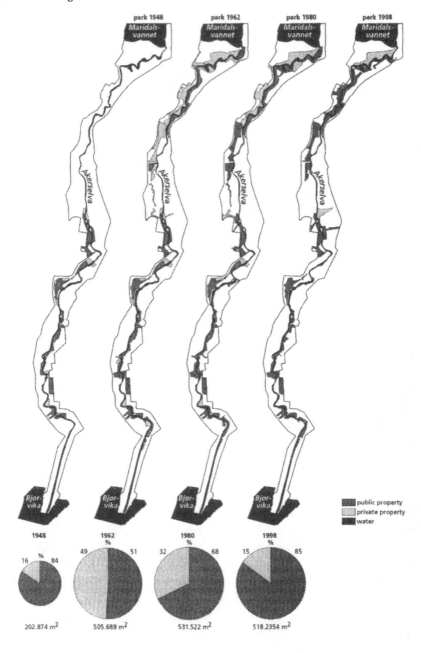

Figure 12.2 Development in public and private park area

planners to follow through and realise infrastructure projects of this kind. Radical anti-establishment movements had an effect on city planning, as they did in other sectors throughout society. The larger projects were effectively stopped by demonstrations, mobilisations of tenants and small shop owners and by squatting. These signs of the times were similar in many cities all over Western Europe.

This led to a change in the planning strategies. Instead of large-scale infrastructure projects, there was a shift towards smaller projects. These were projects that had matured politically and achieved a high degree of consensus, and were carried out step by step – sometimes in very small steps. In other words, there was a change towards incremental planning.

One large project carried out in the 1970s was the City Renewal Project (Byfornyelsen). The old sections of the centre of Oslo, dominated by working-class housing from the 1890s, was by then showing clear signs of decay and even turning into slums. The decline in the manufacturing industries in the centre of Oslo and alongside the Akerselva River Basin had already begun. Rent control kept the rental revenues too low to make it profitable for landlords to invest in upkeep. The political climate changed dramatically. It was then the strong forces who were advocates of investing large sums of public money in upgrading low-rent housing. Their standpoint was largely influenced by similar thoughts in neighbouring countries like Sweden and Denmark. The idea was to invest district by district, taking into consideration that the public did not have full legal and financial control of the areas in question. Co-operation with both tenants and private owners was necessary in most cases. The project ran into severe difficulties when the costs got out of control. Many of the renewed apartments became too expensive for the tenants afterwards. Since they could not afford to live there anymore, many moved out or became dependent on social benefits.

Many of the same people who later became involved in the Akerselva Environmental Park Project had played key roles in the City Renewal Project. They brought with them valuable experience from this project. Because of a fairly high job mobility within the public sector, one could soon find colleagues from the same planning department in the City of Oslo in different important positions in the new state institutions. The Ministry of Environment (established in 1975) was one of these, and the network between these bureaucrats in the City of Oslo and in the Ministry was (and is) strong, rather informal, and effective. They shared the same ideas, the same values, some were even of the same age and some of their children were in the same kindergarten.

By the middle of the 1980s, the City Renewal Project had ground to a halt, due to financial problems and its unpopularity. A new political wave – liberalism – had come to dominate the scene in Norway, as in the rest of Western Europe and North America. Planning in general was clearly put on the defensive.

At about the same time, signs of a transition from the fordist production era to the post-fordist era had become clearly evident in Oslo, particularly in the Akerselva River Basin area. The decline had started in the 1970s, but it accelerated through the early years of the 1980s. Other sectors were undergoing strong growth, and the transition to the service sector was rapidly taking place. Many were concerned that the old factories and the most important monuments of the industrial revolution in Norway would vanish and be replaced by glass and steel 'yuppie' building for lawyers, consultants and PR people.

By this time, the environmental movement had grown strong in Norway. The Gro Harlem Brundtland Commission (Mrs. Brundtland is a former Prime Minister in Norway) had delivered its report in 1986. And many preservation and protection projects had already taken place all over the country.

Some bureaucrats within the planning network decided to launch a new kind of environmental project, a project for the City of Oslo. It is said that the Akerselva Environmental Park started with a gift of NOK 500,000 (80,000 Euro) from the private company Bjølsen Valsemølle – an active mill located near the river – to the City of Oslo. The gift was supposed to be used for environmental or preservation purposes in the Akerselva River Basin. The money triggered off a flurry of ideas in the planning community within the City of Oslo. The NOK 500,000 would not last long and the City of Oslo had little money available for that kind of project at the time. It was therefore necessary to mobilise the planners' network in both the city and the state system. Some of the same planners who had been (or were still) involved in the City Renewal Project decided they would try to get more money from the central government. The idea was presented to the Minister of Environment, Mrs. Sissel Rønbeck (Labour), who immediately liked the project. Though it was not her original idea, she very soon pursued it with all her personality, prestige, and philosophy on holistic planning and politics, but also with state money.

Her original plan was to create a park which would also serve as a monument for the working-class women who had worked hard, long shifts of 12 – 14 hours in the first factories in the Akerselva River Basin. From her point of view, the project should try to preserve the industrial culture of the area. In November 1986, she invited the departments of her own ministry and of the City of Oslo to a large meeting. She presented the idea of creating Akerselva National Park – after a similar model of the national parks that had been or were being established in Norway at the time. The definition of a national park was important because it had certain legal implications. She soon ran into formal problems with this strategy, since national parks had to – by their legal definition – be outside urban areas. She then changed the name to 'Akerselva Environmental Park'. The concept of 'Environmental Park' was new and had weaker environmental and legal implications. It clearly represented an institutional innovation at the time, and Mrs. Rønbeck was very much the driving force behind it.

The Directorate for Cultural Heritage (Riksantikvaren) had allocated NOK 3,250,000 (about 350,000 Euro) in the budget for 1987. For a period of years, the Directorate had been concerned with the decay of the old factory buildings in the area. By that time, they had already done substantial work as far as mapping and getting an overview of the most historical sites in the area. In late fall of 1986, when the Environmental Park initiative came from the Ministry of Environment, they had a 'good position' and a clear strategy for how they should act.

The Akerselva Environmental Park Project

In December 1986, a large press conference was held, and the project was officially launched. It was clear from the start that the project should formally and administratively be based in the planning system of the City of Oslo. By Norwegian planning law, the municipality has the right (in most cases) to regulate physical planning. The state level can – by invoking laws for the protection of the environment and preservation of the historical and cultural heritage – exert strong influence on planning, however. This is most likely to happen if they want to stop projects undertaken by private or municipal actors.

The project was therefore organised with high level decision makers in the city of Oslo and people at similar positions in the Ministry of Environment and the Directorate of Cultural Heritage. The steering committee was led by Ulf F. Beichmann, the technical alderman of the City of Oslo at the time Mr. Beichmann, who is now retired, was highly respected and had great influence within Oslo's planning- and administrative system. Moreover, he had the capability to run and finish large projects. Putting him at the head also made it very clear that Oslo was in charge of the project, although the bulk of the finances to carry out the project itself came from the Ministry of Environment. This a was deliberate move by the ministry, which wanted to contribute to the work already done by the City of Oslo in the Akerselva River Basin area back in the days of Mr. Røhne and up to the present. At the same time, the ministry wanted to make sure that the City would take charge of the project after it was finished.

It should be emphasised that there were no *private* actors or interest groups in the steering committee. It is not easy to get a clear explanation from the participants, but it seems like the experience from the 1970s and the numerous 'involvement-projects' had somehow had a negative effect. It was more or less agreed that such project organisations created more heat than light. It was clearly stated by all the interviewed participants that the strong points of the steering committee were its size and composition. It was not too large, and it consisted of persons who had the power and influence to make things happen in their own organisations. At one point in the project, researchers from the Norwegian Institute for Urban and Regional Research (NIBR) strongly advised reorganising the project and activating the residents of some of the neighbouring districts, who would be the most frequent or most likely users of the Akerselva Environmental Park. These bottom-up initiatives did not survive the steering committee. As representatives of the committee later stated, it was not necessary to bring these interests into the inner circles of the process, because *the people in the committee knew what the public's interests and views were.* Many of the same members were inhabitants and users of the area themselves, and they had had many meetings and occasions where the issues had been discussed in general. However representatives of the steering committee had held several meetings with individual private property owners along both sides of the river. Those meetings were in either to inform the parties about the project and its aims, or to negotiate directly on concrete questions that had to be solved in order to create the Environmental Park. It made no sense to turn these meetings into large public sessions for people whose interests were only indirect. It is interesting that some of the same actors who participated in the steering committee then, and who also participates in similar committees in large urban renewal projects today, are criticising these projects for having 'too large steering committees'.

The steering committee hired Ola Bettum – from In By, a consultancy firm – as one of the main secretaries. Bettum had been working at the Ministry of Environment when the project was developed, and he belonged to the network of public (and semi-public/private) planners. The role of the project secretary cannot be overestimated in a project like this. Mr. Bettum's position was that of a 'neutral' player (independent of any of the public agencies). But he was highly pro-active, trying to create solutions by negotiating with the different fractions and interests represented within the group. At the same time, Bettum – trained as a landscape architect – had the necessary multidisciplinary approach and social intelligence to understand the different 'planning cultures' and motives. In close co-operation with Mr. Beichmann, he was consequently able to create a very efficient steering committee, which managed to attract resources to the project, create common goals, and follow up on the agenda. The direct engagement and interest of the Minister of Environment, Mrs. Rønbeck, also helped and was inspiring to the members.

The main aim of the project was as follows:

"Akerselva Environmental Park should be developed as a national and local area for recreation and learning. It should provide a rich menu of opportunities and be a highly accessible, continuous park belt from the Maridalsvannet (a large lake to the north) to Vaterland (the fjord in the south). The park should at all seasons represent a city park, both for the neighbouring districts of the city as for the city as a whole.

Akerselva shall – with its clean water and well-kept riverbanks, represent a smile in the face of the city. There shall be fish in all sections of the river, and the natural habitat shall be secured.

The important historical and cultural heritage, mainly represented by the pre-manufacturing and industrial historical buildings related to the river, shall be preserved. Further development of the area should build on these traditions and contribute with further qualities. Old and new buildings shall, together with the landscape and the vegetation, create harmonious and changing green but urban space around the river" (Translated from the Kommunedelplan, Oslo Kommune, 1989).

One should notice that the main aims were mostly 'green'. That is, they were focused on environmental and recreational values, learning and preserving the historical heritage, etc. There is not one word about stimulating or restructuring industrial development. It was, however, stated in later interviews that the members of the steering committee clearly realised that these aims had to be achieved within a context of an existing dynamic industrial structure. Several members of the steering committee argued that it was never the intent to create a 'museum' or 'just a recreational park'. It was expected that an environmental park could arise within an active industrial area. It is tempting to interpret this as a rationalisation of the actual process in hindsight.

On the other hand, the planners were, of course, very much aware of the role of private industrial interests. Indeed it is stated clearly that the environmental park should play a secondary legal role in the area of Nydalen, the most densely industrialised area of the old Spigerverket, the old steel mill in the upper northern part of the park. During the 1980s, it was being transformed by a private real estate developer - Avantor - to accommodate modern service industries. Avantor has invested both in the old buildings and in new ones and has contributed significantly to the public infrastructure in the area. For this particular area, a special city plan (Kommunedelplan for Nydalen) had been produced. That plan had regulations similar to those in the Akerselva Environmental Park plan, but it was more detailed in many cases. Nevertheless, it was fairly 'easy' on industry, allowing an intensive use of the available land, which is often necessary to make it profitable for private investors. Interestingly enough, the managing director of Avantor, Mr. Joyce, had previously worked at the planning departments in the City of Oslo. In a way, he belonged to the same network. In this sense, he had the same professional language, some of the same values, and of course many of the same connections as the public planners. This has probably made communication between the private and the public actors fairly easy and efficient. However, Avantor never had a formal position in the project.

The project was only supposed to last two years, 1987 and 1988. The initial budget was NOK six million (750,000 Euro), though this just represents the visible state contribution. Early on, the organisers were able to get more resources for the project as it developed. The steering committee soon decided that they needed more time, and the project was prolonged by two years (1989 and 1990). The amount of state money allocated to the project was increased. The NOK six million was raised to 23 million krone (about 3.5 million Euro) and was supplemented by regular investments through the annual city budgets. The supplements must have amounted to much larger sums, but it is difficult to make good estimates. As we have seen, the purchase of property started in the early 1920s and continued through the 1990s. It has not been possible to make an overall evaluation of the total value of the public property investments in this project.

Most of project money went into the preservation of historical and cultural landmarks. More than NOK 10 million went into this part of the project, mostly to concrete preservation projects. The Division of Conservation (Byantikvaren) was generally quite creative in finding other sources of public funding. In 1988-1989 Norway experienced a sharp recession due to the drop in oil prices in 1986-1987. Unemployment was rising fast and the sector that was hit hardest was the construction industry. This meant that many architects lost their jobs and recent graduates had a hard time finding their first job. The Division of Conservation managed to get resources through public unemployment programmes in order to run special projects (33) targeted for the Akerselva Environmental

Park. There, 19 architects and planners did detailed mapping of individual cultural and historical buildings and sites as a part of their training. These plans were eventually turned into official plans. This gave the cultural heritage fraction a stronger position, which it has maintained after the project officially ended.

The Results of the Akerselva Environmental Park Project

What were then the results of the project? It is useful to differentiate between the *intended* results and the *unintended* ones. The intended results may be summarised as follows:

- it was a dynamic project in the sense that it was possible to refine the aims and increase the level of co-operation as the project developed;
- the public plan was achieved, which in turn created a stronger awareness of the natural, historical and cultural values of the Akerselva River Basin;
- through the plan, which gave the city formal tools, but also through the dialogue with the private owners, the project created a higher degree of awareness of the qualities of the area. This has, in turn, changed the private sector's attitude toward the park. It is now considered a valuable resource and one of the main location factors;
- people's awareness of the park has been increased. It has become a popular recreational area and one of the major 'green' transit corridors in the city;
- the project has created one of the most important park and recreational areas in Oslo;
- the project has created one of the most important historic industrial heritage parks in Norway;
- the Rønbeck model – bringing together various state, regional and local authorities and the private sector – has later been used in many other similar projects throughout Norway with good results.

It is important to notice that many of the effects did not appear during the formal project period. Some are becoming evident as time goes by and as public and private investments in the area add to the total picture.

The most interesting outcomes may have been *unintended*. There has been an unusually positive effect on industrial regeneration in this part of Oslo. In fact, the industrial transformation that has taken place in the area over the last 15 years is remarkable. During the peak years of the Spigerverket steel mill, the area had 2000 blue-collar jobs. Almost all these jobs were lost in a few years time. The area bounced back during the 1980s and 1990s. At present, the area has more than 5000 white-collar, highly skilled, highly paid, low-polluting jobs in the service sector and information technology industry.

The transition has been remarkable in terms of its size and speed, but also in the sense that it has received so little public attention. Industrial regeneration projects elsewhere in Norway get major public support from regional policy programmes and draw scientific attention before, during and after the projects have started (through evaluations). This project, however, is hardly known to researchers, planners or industrial policy-makers. The explanation has not been investigated systematically in this project. But spokesmen for Avantor who are in day-to-day contact with its tenants – the firms – suggests several reasons for the success of the project. One of reasons that Avantor gives is that they can provide a much faster service to firms who want to rent or invest in the area, faster than in other parts of Oslo where these responsibilities have not been privatised.

Another reason is that the location factors are right for the new industries. The site is near the important customers, it has a fairly efficient road system and will have (it is planned but not yet financed) a new metro line with a station in the area, it is near the large universities and R and D institutions, it is near the large skilled labour markets, and it is near other firms and subcontractors. In a short time, and without any deliberate public planning, the area has developed several *industrial clusters* (Porter, 1990) and turned into an *industrial milieu* (Marshall, 1891). There seems to be a critical mass of firms, and that level of economic activity generates further growth and attracts other firms.

Its central location in the city and the short distance between the tenants in the Nydalen area makes it efficient, with respect to time and cost, to be located here. It is within a short walking and biking distance of other business relations within the area. And, not to be underestimated, it is considered a pleasant area to work in. The Akerselva Environmental Park has improved the quality of life in the area as well as in the neighbouring urban settlements. Consequently, it is possible to enjoy the park itself for recreation (lunch breaks, jogging, cycling). Or it can be used as a green transit corridor to the forest belt around Oslo to the north or to the fjord or the centre of the city to the south. People can also use the trail system to get to restaurants, cafes and cultural services in the neighbouring districts on both sides of the river and especially in Grünerløkka, an old part of the city dating back to the 1890s which was renewed and became very popular in the same period.

Conclusions

There are three important lessons to be drawn from this case. The first one is obvious: planning, government and governance always take place within a wider framework of economic and technological change. Planning and governance are always easier if they go with and not against ebb and the flow of the market. The Akerselva Environmental Project started off as a 'green' idea. The planners who came up with a consensus strategy to achieve their non-industrial aims ended up promoting the industrial regeneration of Oslo's economy. They were thus (unintentionally) ahead of most economic and industrial planners at the time, who were still working with old theoretical paradigms. The Akerselva Environmental Park just happened to stimulate the new forces of economic growth. One optimistic conclusion is that the internal contradictions between economic growth and sustainable development seem to be less dominant in post-fordist society.

The second lesson to be drawn concerns governance. The Akerselva Environmental Park has been presented within the theoretical framework of this book, which describes urban governance as a bottom-up process characterised by the participation of many autonomous agents, public-private partnership, lack of common goals etc. The investigation shows, however, that one of the main causes for the positive effects and the efficiency of the project has been the large share of public ownership of space in the park, before, during and after the project. In this sense the project has been a rather typical social democratic project conducted along traditional lines.

It is, however, also a case of network governance, although the network is mostly related to smaller circles of public officials in different layers of government. Some would say that this is a typical characteristic of a social democratic project. This brings us to the third and perhaps least welcome lesson, as far as democracy is concerned. The record shows that one of the reasons for the success of the project is governance through bureaucratic elitism within a rather small and partially-closed network. Akerselva Environmental Park is not a project where the masses were involved in bottom-up participation. Rather, the project was run by a handful of bureaucrats who work well with each other, who share the same values, who have pretty much the same professional background, and who know how to extract resources from different public sources. If this is a worrisome development, it is only because large segments of the population are not taking part in the planning process. On a positive note, networks of this kind seem to get the job done. In this case, the results seem to have been beneficial to the business community and the public interest at the same time.

Notes

[1] Special thanks are due to Fred Olav Sørensen who first supported the project and to Karl Otto Ellefsen and Aasne Haug of the Oslo School of Architecture who did the physical mapping and produced the maps for this chapter.

References

Kann, F. and Halvorsen K. (1995), *Næringslivets Utviklingspotensialer i Osloregionen. Modul A: Strukturanalysen.* NIBR notat 1995/105: Oslo.

Luhmann, N. (1982), *The Differentiation of Society,* Colombia University Press, New York.

Marshall, A. (1891), *Principles of Economics,* McMillian, London.

Oslo Kommune (1989), *Forslag til Kommunedelplan. Akerselva Miljøpark,* The municipality of Oslo, Oslo.

Porter, M. (1990), *The Competitive Advantage of Nations,* The Free Press, New York.

Rhodes, R. A. W. (1997), Foreword, in W. Kickert, E.-H. Klijn, J.M. Koppenjan and F.M. Joop (Eds.), *Managing Complex Networks. Strategies for the Public Sector*, Sage Publications, London.

Schumpeter, J. A. (1943), *Capitalism, Socialism and Democracy*, Harper Torch Books, New York.

13 Urban Governance and Infrastructure: Coping with Diversity, Complexity and Uncertainty

MARTIN DIJST AND WALTER SCHENKEL

Growing Diversity, Complexity and Uncertainty

As a consequence of economic, technological and socio-cultural megatrends, the economic and social structure of urban society has changed and is still in motion. The internationalisation and informalisation of economies has changed the competitiveness of urban economies and powered the replacement of industrial economic functions by the service sector. The 'old' locations of economic functions in the city are often abandoned for new places of business in the suburbs. Also, lifestyles have evolved over the past decades, becoming less and less uniform. With the help of the private car, people's action spaces have expanded, becoming more differentiated among individuals. These changes are accompanied by serious economic, social and ecological problems which threaten the competitiveness and sustainability of cities.

The urban transformation is characterised by a growth in scale and complexity of the economic, social and political networks in which actors take part. The transformation is accompanied by a diversification in the mobility patterns of persons, goods and information. The diversity, complexity and dynamics of the network society lead to uncertainty among those who have to manage and control the urban transformation processes. Their task is to improve urban performance in economic, social and ecological respects. Their uncertainty is fed by a lack of knowledge about the interrelations between economic, technological and socio-cultural megatrends, about the behaviour of different actors, and about ways to analyse and influence these linkages.

In this book, we focus on *urban infrastructures* (such as communication and transport, culture and recreation, and environmental infrastructures). Especially in the transport and communication sector, infrastructural developments lead to spatial distribution patterns, economic structures and social behaviours which require new forms of urban governance. The main objective of this book is to develop concepts, methods and strategies for investments in urban infrastructures in order to govern increasing diversity, complexity and uncertainty, features which characterise the urban transformation processes. It is shown that different sets of tools affect the capacity to manage urban transformations. Moreover, these processes are shown to be interrelated. Research work by COST-CIVITAS members has been based on a dual approach to urban infrastructures and development: a *functional* and a *governance/management perspective*. As a result this volume is a rich collection of theoretical contributions and case studies (see chapter 1, section 'Outline of the Book').

These contributions offer some support to decision-makers. Knowledge and information should reach actors – whether engaged in daily life or in the planning practice – in order to improve urban governance, co-operation and decisions on joint urban planning. From a functional perspective, planning models may indicate where planning strategies and transport systems are insufficient. However, concrete conclusions have to be made by people who are able to translate scientific results into political demands. From a governance/management perspective, network analysis is able to improve the information that underpins existing arrangements among actors and can recommend improvements. Nonetheless, efficient management and network building is highly dependent on the willingness of administrations, landowners, businesses, politicians, and societal groupings to take part in such a process, to exchange information, and to accept uncertainties.

In this concluding chapter, we discuss how these two perspectives on urban infrastructures and urban performance could be linked to each other. For this purpose, the next section proposes a framework for an integrated approach. Section 'Learning from theoretical and practical experiences' combines the most important issues raised in the contributions to this book. In the concluding section, we formulate some suggestions for research in the near future.

Framework for Effective and Efficient Urban Governance

The main aim of urban policies is to *influence urban performance*. Local authorities cannot *directly* influence the behaviour of urban actors in order to improve performance. They can only influence this behaviour in an *indirect* way by changing the *conditions* in which behavioural decisions are taken by the actors. The effectiveness of the policies to change these conditions is dependent on an understanding of the relations between processes in society at large, the conditions for the behaviour of urban actors, and the development of the urban structure in the course of time.

Conditions from the Functional Perspective

A prerequisite for effective and efficient urban policies is to have a good understanding at the micro level of the behaviour of different actors in daily life within the urban system (the functional perspective), while taking into account the processes in society at large at the macro level. Together with the existing urban structure and the urban performance at that time (T_0), these processes will influence the set of conditions for the behavioural decisions on the micro level (Figure 13.1).

This set of conditions is also called 'accessibility'. These opportunities to travel will influence the locational decisions of households (where they live and where they work) as well as the locational decisions of companies and employees. Also, the daily travel decisions, like mode choice and travel routes, are influenced by this set of conditions. The aggregation of all these individual decisions has consequences of the performance of the city and the changes which are taking place with respect to the land use and supply of transport systems at time T_1.

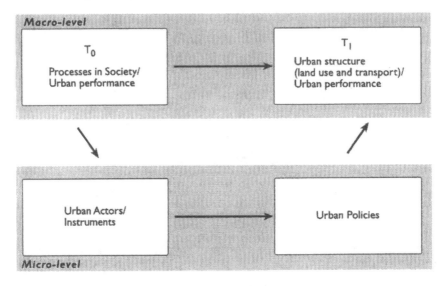

Figure 13.1 **Relations between macro and micro level from a functional perspective**

Conditions from the governance/management perspective To gain a better understanding of the type of governance/management best suited to influence the behaviour of different actors in daily life, we need to know how the behaviour of the urban actors in the planning process is linked to developments in the urban system (the governance/management perspective). The conditions for decision-making and implementation of these urban actors are influenced by the economic, social, cultural, spatial, political and administrative processes in society at large and the urban performance at a given point in time, T_0 (Figure 13.2). These conditions refer to the type and number of actors involved in the planning process and the available instruments.

The participating actors and the available instruments will influence decision-making processes. These decisions will result in the formulation and implementation of urban policies and investment programmes. Subsequently, the use of urban structure and/or land uses and the supply of transport systems will be changed. That, in turn, will affect urban performance at time T_1.

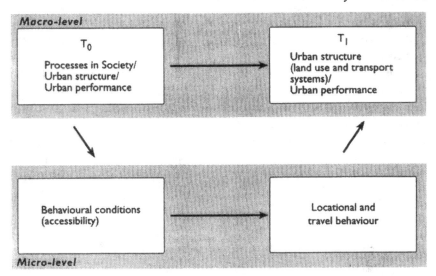

Figure 13.2   Relations between the macro and micro level from the governance/management perspective

Interdependence of Both Perspectives   Both the functional and the governance/management perspectives are linked to and dependent upon each other. Figure 13.3 shows these interactions. The increasing diversity and complexity in daily life means that new actor groups are growing in importance and power, while others are

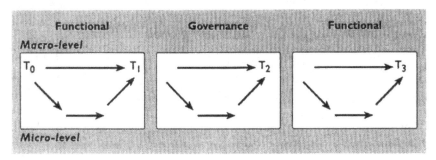

Figure 13.3   Interactions between functional and governance processes

losing their influence. These developments will change the conditions in the planning process, which can have consequences for future policy decisions.

On the other hand, the development of actor networks in the planning process, the selection of instruments, and the formulation and implementation of urban policies will change the land uses and supply of transport systems as well as urban performance. These changes within the urban system will modify the behavioural conditions for households, companies and institutions in daily life. In time, this can have consequences for their locational and travel behaviour.

Learning from Theoretical and Practical Experiences

Functional Approach

It has been demonstrated, how the rapid increase of mobility has led to new patterns of consumption of space and time. Urban sprawl in terms of housing, shopping facilities and workplaces is one result; flexible working and shopping times and extended recreation are another. Within this context, *transport planning* has to consider more complex travel patterns. Furthermore, it has to deal with both the hardware (investment) side of infrastructure, as well as the software (operational handling, pricing) side.

The paper by *Dijst, Jayet* and *Thomas* represents a first step towards a deeper understanding of the complex relations between a series of different and related factors regarding urban performance. The authors make an important distinction between accessibility of the individual or household and accessibility of an activity place or urban area. Accessibility is considered a key issue in transportation and urban policies: both determine territorial transformation. As these authors demonstrate, urban governance should consider two dimensions of trade-offs: between individual transportation costs and people's travel needs; and between collective transportation costs and production costs of companies. The results of these trade-offs are reflected in the land values or bid rents. To sum up: the link between location decisions (by firms or households) and the economic, social and environmental performance of the city is extremely important and should be guided by spatial configurations and transportation systems.

Boffi and *Nuvolati* take up a theme which is rather new within urban policies: the time factor as one of the main features managing the transformation of the city. They argue that planning approaches should be able to take the strong fragmentation of time into which is a consequence of the combination of labour, consumption and social activities by each individual. Their approach has potential for the future, since there is growing interest in lifestyles and time uses as a crucial variable for spatial planning and urban governance. In their case study of Milan, they underline how the mix of mobility patterns is related to different population groups using the city. Four innovations are conceptualised: the integrated policy design concerning public services and timing; the relevance of time in cabling the city; the time-bank improvement; and the impact of time flexibility on urban mobility. In fact, the use of time, considered a scarce resource, depends on the spatial organisation of activities. In the future, time policies should be developed not only by considering the number and type of actors, but also by delimiting metropolitan areas in order to design planning strategies related to different mobility profiles. In this context, the article of *Riganti* points out the impact the two Crossrail projects in Milan and Turin will have on the organisation of the region, due to the increase in the levels of accessibility and the possibility of a further urban expansion. Nevertheless, the shift from private to public transport is not directly expected as an effect of the density of existing networks and mobility opportunities. Here, individual and collective time patterns should not be underestimated. However, cities see new Crossrail systems as an opportunity to govern change in a more general framework.

Kanaroglou and *Scott's* paper offers some support for the application of transport models. As they point out, considerable empirical evidence is available to document the relationship between transportation and land-use policy. Yet more often than not, policy is formulated and practised independently from land-use considerations. Integrated urban models can be effective in offering insights into the effects of urban policies on urban performance. Since most actors are interested in real estate values, it is to be hoped that such models will get linked to their actors' interests.

Verhetsel's paper may be seen in this context. She describes the implementation of a traffic model to evaluate the impact of different policies on urban traffic, underlining the different impacts of planning, infrastructure, regulatory and financial measures. She does not expect too much from planning measures. They may, for instance, stimulate people to make use of public transport, but regulatory and financial interventions would still have stronger effects. Planners and geographers ought to conduct more research into the local spatial effects of other non-physical measures (such as network management and financial incentives). However, the Flemish model presented in the paper has the potential to be used in an argumentative discourse on sustainable urban development.

Governance/Management Approach

Research reveals some new forms of urban governance. Networking and negotiation between a variety of actors in urban development are now recognised as an important complement of or even as a substitute for regulatory measures by local or regional governments. Urban governance implies organisational capacities and interaction between public and private actors. Individual power is no longer defined by objective financial resources and hierarchies but by actors' positions within the network, their relationships to each other, and the functioning of the network itself. City governance, thus, is assumed to have adopted a new approach to decision-making and management. The general picture – at least in democratic political systems – is one of increasing differentiation of responsibilities for managing 'new' complex problems.

Mennola presents a useful framework to compare single case studies with a general understanding of relevant characteristics of European cities. He distinguishes four ideal types of city governance – the unitary city, the village city, the multi-level city, and the private city – along four dimensions of interaction: the socio-economic area, the political community, the public administration, and the legal framework. He initiates a dialogue on improving the general comparative framework. *Schenkel's* theoretical paper on policy networks can be understood as a specification of one of the four types of city governance. Network management is assumed to be close to the multi-level city. Emphasising the shift from the bureaucratic to the network state, he demonstrates how urban policy can be improved if it is supported by communicative steering and network management. The main point is to focus on the actors rather then on the physical urban problems.

The two case studies by *Güller and Schenkel* and *Halvorsen* are good examples of the benefits of building up relevant actor networks. Compared to the traditional ex ante regulation of land use, case studies like these, focusing on the importance of network management, suggest that agreement-based regulations could lead to more successful planning and implementation. Although large infrastructure programmes are not going to disappear totally, it is shown that significant changes can be made in a city by gradually improving infrastructure without even having a clear idea of the ultimate goal. One target could be economic, social and environmental sustainability, but only if it is recognised as a normative, uncertain and process-oriented concept. Both case studies are good examples of new trends in urban planning, combining site potential, human resources, and management methods to move toward win-win situations.

Some of the obstacles should not be underestimated, however the Akerselva project was initially based on public ownership of space and public purchase of land, and the networks were built up mostly among public planners. Later, the project turned into a public-private partnership. From the beginning, the Zurich project was highly dependent on the quality of collaboration between the public administration and the private landowners; most land in Zurich is private property.

However, business prefers land markets with minimum constraints, regulations with maximum flexibility and certainty of outcome, and vigorous public funding of infrastructure. The network approach clearly supports the strategy towards public-private partnerships. The network approach clearly supports the strategy towards public-private partnerships both in terms of decision making on stepwise development, as well as financing on the base of iterated returns on investment.

The Geneva experience, as presented by *Stein*, analyses changes in the behaviour of certain groups among the population when they use public space. The example of public space demonstrates how top-down approaches need to be supported, sometimes even replaced, by bottom-up attitudes about managing the urban environment. Public space is considered as one of the major factors for the re-composition of the urban environment. In contrast to the Zurich case (no inhabitants yet), the Geneva case was characterised by working-class housing and now has a high density of population. Therefore, Stein supposes that a successful project should be embraced by the various actors involved, in particular the inhabitants and users. *Lami*, on the other hand, stresses the importance of financial aspects and related feasibility evaluation techniques for transportation investments. Their application allows the necessary interaction between plan and market. The analysis of financial data makes it possible to specify subject matter, parties involved, costs and relative coverage. Despite the commonly shared view, the chance to use infrastructural policies to increase the value of urban areas has largely been overestimated. The most effective way to derive some multiplier effects from the advantages generated by public investments in transport has been linked to partnerships based on local building projects.

Combined Approach

As mentioned before a close link exists between the functional and the management approach. The increased world-wide mobility of goods and persons, labour and capital has contributed, in Western societies, to a fundamental transition – namely from an industrial to a service-oriented economy. This change involves decades of turbulence and uncertainty on labour and real estate markets. *Cities are on the move*. In light of the uncertainties and the new challenges regarding the distribution of advantages and disadvantages among the different strata of the population and economy there is a need to reconsider how cities (the crucibles of change) are governed. Markets and networks are not meant to replace or avoid clear regulation. Rather, they are supposed to integrate a wide range of interests and to increase the responsibility of each actor for the others.

It should be kept in mind that municipal governments deal not only with their internal problems. They are also faced with increased interurban and international competition. In part, the efficiency of urban governance is a decisive factor in this competitive climate. But the level of infrastructural equipment is of relevance too. Municipal and regional governments are well aware that only cities with a substantial number of taxpayers are able to develop their infrastructures in a competitive way and to provide for the necessary social services. Keeping these taxpayers in the city, or bringing them back, has become one of the key tasks of urban development and urban governance.

What both perspectives have in common is that traditional practice has to be reviewed. Traditional planning and governing is no longer sufficient if it is reduced to plans, hierarchical structures, and an elite group of planners, administrators, and landowners. Models have to be tested with respect to their political practicability and feasibility in life. Urban governance should not be reduced to the enforcement of legislation. It increasingly involves the management of instruments and actor networks on a strategic level. On an operational level, urban governance calls for the participation of a wide variety of actors in the phases of planning and decision-making, financing, implementation and land use. To that end, urban governance can be improved along seven dimensions:

- renewal and development of urban identities;
- replacement of plans by strategies and discourse;
- building up confidence;
- building networks and coalitions;
- setting interim targets (as in a milestone policy, i.e. stepwise, but with clear deadlines);
- human resource management (i.e. promotional spirit, leadership, knowledge, and capacity-building networks);
- promotion of sustainability as a general guideline. These points should be considered in the 'traditional' infrastructure or land-use planning as well as in urban governance, understood as network management and network building.

Governing Cities on the Move: a Research Agenda

It is our thesis that functional and governance perspectives should be more integrated with each other in order to govern diversity, complexity and uncertainty. This book contains some theoretical and practical building blocks that we can use to build an integrative framework. But this volume is just the first step in that direction. Several questions have to be answered before we can take the next steps. Two questions can be formulated:

1. *In what way are the functional and governance perspectives on urban performance linked to each other in practice, and how could the integration of both perspectives be improved?*

We need more information on the way in which knowledge and information on the behaviour of actors in daily life can be produced and used in planning processes. Growing diversity, complexity and uncertainty make it necessary to change our concepts and analytical tools to increase this knowledge and thereby improve the effectiveness of urban governance. A systematic review of the types of concepts, methods and models which are used to 'measure' this behaviour is still lacking. We also need to know more about the influence of various factors on the use of knowledge and information.

2. *In what way could the moral dimension of actors in a policy network be improved?*

The increasing diversity, complexity and uncertainty can be governed by using regulations such as laws. This approach can increase the burden of the control and administrative functions in the city. We should ask ourselves the question if further regulation could be avoided in order to integrate the interests of all potential actors in the planning process. By increasing the responsibility of each actor for the interests of the other actors, this goal could be reached.

We think that more research is needed to integrate the functional and governance perspectives. At the same time, we should not forget that science is no longer able to resolve all the uncertainties and doubts that might arise. Both scientists and practitioners must learn to deal with complex problems in a longer time frame. There are hardly any immediate solutions, but if scientists start taking part in practical planning processes, they will be in a better position to suggest ways to manage these uncertainties. A concerted effort by all actors in the policy network can contribute to the development of visions. This will give practitioners a basis on which to evaluate every step taken towards these goals.

Printed and bound by PG in the USA

USA2019PGIL